Sliding Mode Control
of
Switching Power
Converters

Techniques and Implementation

Sliding Mode Control
of
Switching Power Converters

Techniques and Implementation

Siew-Chong Tan
Yuk-Ming Lai
Chi Kong Tse

CRC Press
Taylor & Francis Group
Boca Raton London New York

CRC Press is an imprint of the
Taylor & Francis Group, an **informa** business

CRC Press
Taylor & Francis Group
6000 Broken Sound Parkway NW, Suite 300
Boca Raton, FL 33487-2742

First issued in paperback 2017

© 2012 by Taylor & Francis Group, LLC
CRC Press is an imprint of Taylor & Francis Group, an Informa business

No claim to original U.S. Government works

ISBN-13: 978-1-4398-3025-3 (hbk)
ISBN-13: 978-1-138-07549-8 (pbk)

Visit the Taylor & Francis Web site at
http://www.taylorandfrancis.com

and the CRC Press Web site at
http://www.crcpress.com

Contents

8 Design and Implementation of PWM-Based Sliding Mode Controllers for Power Converters 167

Preface

As the characteristics of power sources and electrical/electronic loads become more widely varied, nonlinear, and unpredictable, the control of the power converters that provide the necessary power processing functions will play a crucial role in optimizing performance and maintaining the needed robustness under various operating conditions. Conventional control approaches based on small-signal linear techniques are found to be incapable of achieving the necessary regulation, dynamic response, and stability requirements needed for these systems. Much research effort has been devoted to the development of advanced control methods that are able to address the complex requirements of power conversion systems. In particular, modern control theories such as sliding mode control, fuzzy control, adaptive control, etc., have been applied to the control of these systems, and their feasibility has been studied. Among them, sliding mode control has been most widely investigated and has demonstrated to be a highly promising solution for both current and future generations of power converters.

Following the trend of this development and taking advantage of the mature theoretical framework already being laid out on the subject, it is timely to present a complete exposition of the development of sliding mode controllers for power converters. This book presents an in-depth and thorough account of how such kind of controllers can be practically engineered to suit the purpose of controlling power converters. The dissemination of such knowledge is now timely and necessary, especially since the electronic industry is moving toward using renewable energy sources and widely varying loads, which can only be adequately supported by power converters using nonlinear controllers.

We developed this book with the following objectives in mind. First, we aim to offer a comprehensive overview of the principles and methods in the application of sliding mode control to power converter systems for the general readership. Second, for the more advanced readers, we aim to provide a systematic exposition of the mathematical machineries and design principles relevant to the construction of sliding mode controllers, and building upon that, to introduce and impart new practical approaches of designing such controllers. Our third objective is to demonstrate the practical implementation of sliding mode controllers based on analog circuits and to present their supporting design rules. Our final objective is to promote an appreciation of nonlinear control in general by presenting it from a practical perspective and using terminology that is familiar to the engineers.

In a nutshell, *Sliding Mode Control of Power Converters: Techniques and*

Implementation will be a guide to understanding the sliding mode control principle, its application to power converters, and the practical realization of sliding mode controllers. By combining theory with application and relating mathematical concepts and models to their industrial targets, this book will be equally accessible to readers with analog circuit design, power electronics, or control engineering backgrounds. We believe that this book will be of interest to students and professionals alike in the field of electrical and electronic engineering. At the same time, we believe that our approach to the modeling and implementation of sliding mode controllers for power converters will help power electronics and IC industry professionals design effective and high-performance controllers for power converters.

The book is organized as follows. We will begin in Chapter 1 with a discussion on the basic principles and theory of sliding mode control to familiarize readers with the core terminology and background of sliding mode control. This is followed by Chapter 2, in which we will give a review on the basics of power converters and their control. A short discussion on the common types of control techniques available and the current progress of the research work on the control of power converters are also provided. Next, in Chapter 3, we move on to address the important concepts, operating principles, and properties of the sliding mode controller that are relevant to the application of power converters. Here, we will also provide a detailed review of the state-of-the-art research work and some common practices in the development of sliding mode controllers for power converters.

Then, in Chapter 4, the practical design process of sliding mode controllers based on hysteresis modulation for the power converters is described. In Chapter 5, the problem of switching frequency variation in traditional sliding mode control, due to the deviation of operating conditions, and the effectiveness of applying adaptive control solutions in sliding mode control in alleviating the problem, will be thoroughly covered. Chapter 6 introduces a practical technique for implementing sliding mode control with fixed frequency for power converters operating in continuous conduction mode. Then, the idea of implementing fixed-frequency sliding mode controllers using equivalent control is further explored in Chapter 7. This includes the derivation of the system models and sliding mode control laws for the discontinuous conduction mode converter counterparts. In Chapter 8, we extend our discussion to the design and practical circuit implementation of the pulse-width-modulation-based sliding mode controllers.

Up to this stage of the book, our discussions have been focused on the design and implementation of sliding mode controllers using the output voltage of the converters as the control variable for constructing the sliding manifold. For Chapters 9 to 11, the design and implementation of the sliding mode controllers are extended to cover power converters with non-conventional sliding manifolds that are constructed nonlinearly from the current error and voltage error. In Chapter 9, the design and implementation of sliding mode controllers based on current error and voltage error will be discussed. In Chapter 10, the discussion will be focused on how a reduced-state nonlinear sliding manifold may be used for the control of a high-order converter like the Ćuk converter. In Chapter 11, a non-conventional type of sliding mode controller based on a double-integral sliding surface for improving the steady-state regulation is discussed.

For the completion of this book, we must give our sincere gratitude to a number of people, institutions, and organizations. First, we would like to thank all our friends and colleagues in the Department of Electronic and Information Engineering at Hong Kong Polytechnic University, and the members of the Applied Nonlinear Systems Research Group, for their friendship, encouragement, and inspiration. Special thanks are due to Prof. Luis Martínez-Salamero, Universidad Rovira i Virgili, Tarragona, Spain, who is not only a great friend to us, but is also an important collaborator of some of our research work in this area. Luis is by far the most knowledgeable person in sliding mode control applications in power electronics that the authors personally know. We would also like to take this opportunity to extend our gratitude to Prof. Ashoka Bhat, Dr. Martin Chow, Prof. Adrian Ioinovici, Dr. Joe Liu, Dr. Franki Poon, Dr. Siu Chung Wong, and Prof. Xinbo Ruan, who are all experts in different areas of power electronics, and whom the authors have worked with and learned a great deal from. Next, we would also like to thank the staff of CRC Press, especially Ms. Leong Li Ming and Ms. Amy Blalock, for their professional and enthusiastic support of this project. The completion of this research work would not be possible without the financial support of the Hong Kong Research Grant Council and the Research Committee of Hong Kong Polytechnic University. Last, but not least, we must thank our families for their support and understanding throughout the course of our academic careers.

Finally, it is our pleasure to dedicate this book to all those who have been passionately involved in the work on sliding mode control and the control of power electronics.

Siew-Chong Tan
Yuk-Ming Lai
Chi Kong Tse

1

Introduction to Sliding Mode Control

CONTENTS

1.1 Introduction

The earliest published works that introduced the concept of sliding mode (SM) control can be traced back to the 1930s, and the earliest forms of SM control realization were successfully applied for ship-course control and the control of DC generators [102]. In fact, the development of the theory and applications of SM control were first initiated by Russian engineers, and in the 1950s, the theoretical framework that later facilitated the widespread applications of SM control was reported in the Russian literature [21, 101, 102]. The work was subsequently disseminated outside Russia in English written manuscripts by Itkis (1976) and Utkin (1977) [21]. Since then, the SM control theory has

1

aroused a lot of interests from control theoreticians and practicing engineers around the world.

In simplest terms, the SM control is a kind of nonlinear control which has been developed primarily for the control of variable structure systems [21, 68, 85, 101, 102]. Technically, it consists of a time-varying state-feedback discontinuous control law that switches at a high frequency from one continuous structure to another according to the present position of the state variables in the state space, the objective being to force the dynamics of the system under control to follow exactly what is desired and pre-determined.

The main advantage of a system with SM control characteristics is that it has guaranteed stability and robustness against parameter uncertainties [102]. Moreover, being a control method that has a high degree of flexibility in its design choices, the SM control method is relatively easy to implement as compared to other nonlinear control methods. Such properties make SM control highly suitable for applications in nonlinear systems, accounting for their wide utilization in industrial applications, e.g., electrical drivers, automotive control, furnace control, etc. [21].

In this book, we are concerned with a particular class of variable structure engineering systems, known as power electronics converters. The aim of this introductory chapter is to introduce the basic concepts and mathematical background of SM control that are necessary for understanding the discussions covered in the following chapters.

1.2   General Theory

Let us start by considering a system in a three-dimensional space. Imagine that there exists a plane in this space. On this plane, there is a point O, which we call the equilibrium point. This equilibrium point represents a stable attractor where any trajectory that touches it will settle upon it. It is also a point which we would like to drive the trajectory of our system to.

Next, we consider that the system's controlled trajectory is arbitrarily located in space and is far away from the plane. Without any control action, the trajectory will move according to the natural characteristics of the system. However, when a control action is given, the trajectory can be altered in a "preferred way." The direction in which the trajectory moves is dependent on the type of control action given. A series of different control actions may be given to the system such that regardless of its initial condition, the controlled trajectory will first move toward the plane, and upon reaching the plane, will slide along the plane toward and eventually settle upon O.

A control such as this is known as SM control. The plane which guides the trajectory is called the sliding plane or sliding surface, or more generally, the sliding manifold. The control actions required for performing the SM control

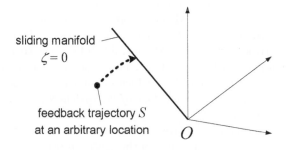

(a) Reaching phase

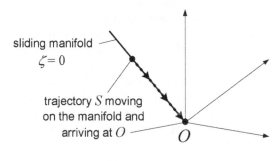

(b) Sliding phase

FIGURE 1.1
Graphical representations of SM control process: (a) Reaching phase—illustrating trajectory S moving toward the sliding manifold irrespective of its initial condition; and (b) Sliding phase—illustrating trajectory S moving on the sliding manifold and stopping at the origin O.

will involve very fast switching between different control functions. The sectors of the space in which the trajectory can be made to perform SM control is called the sliding regimes.

With all these terminologies explained, let us put the definition of SM control in a more formal sense. For any given system, if a sliding regime exists and the sliding manifold $\zeta = 0$ possesses a stable equilibrium point O, when operated in sliding mode, the feedback tracking trajectory S, regardless of its location, will be driven toward the sliding manifold, and upon hitting the manifold, it will induce the control of the system to switch alternately between two or more discrete control functions U_1, U_2, ..., etc., at an infinite frequency, such that the system's trajectory will be trapped precisely on the sliding manifold such that $S = \zeta = 0$, and eventually the trajectory will be directed toward the desired equilibrium point O.

Figure 1.1 gives a graphical representation of the trajectory of a system under SM control. The entire SM operation can be divided into two phases. In the first phase (known as *reaching phase*), regardless of the initial position of the controlled trajectory S, the SM control will force the trajectory toward the sliding manifold (refer to Fig. 1.1(a)). This is possible through the compliance of the so-called *hitting condition*, which *guarantees that, regardless of the initial condition, the controlled trajectory of the system will always be directed toward the sliding manifold* [21].

When the trajectory touches the sliding manifold, the system enters the second phase (known as *sliding phase*) of the control process and is also said to be in SM operation. The system will then be controlled by a series of infinite switching of its control functions such that the trajectory is maintained on the sliding manifold, and is concurrently directed toward the desired equilibrium point O and finally settling at O (see Fig. 1.1(b)). Importantly, by having a control process that reacts only to the way the trajectory behaves, the trajectory will be immune to the effects of the parametric changes and external disturbances. In other words, the control process utilizes the sliding manifold as a reference path, on which the controlled trajectory will track and eventually converge to the origin to achieve steady-state operation without any consideration of the system's parameters and operating conditions. This is possible by satisfying the so-called *existence condition*, which *ensures that the trajectory at locations near the sliding manifold (within a small vicinity of the sliding manifold) will always return to the sliding manifold*, and *stability condition* which *ensures that the trajectory of the system under SM operation will stay on a stable equilibrium point* [21]. Note that it is the stringent enforcement of the trajectory to move along the manifold and stop at the equilibrium point that makes the SM control highly robust against system's uncertainties and environmental changes.

The details of the derivation of the hitting, existence, and stability conditions will be introduced later in this chapter. For now, we would like to reiterate our definition of SM control as *any control form that enforces on its system, the three fundamental mechanisms of the SM process, namely, hitting the sliding surface, staying on the surface, and converging to the stable equilibrium point.*

1.3   Properties of Sliding Motion

1.3.1   An Ideal Control

It should now be clear that the basic principle of SM control is to employ a certain sliding manifold (surface) as a reference path such that the trajectory of the controlled system is directed to the desired equilibrium point. Intuitively, this is possible only because the SM control is inherently adopting an infinite control gain which enables it to trap the trajectory to slide along the manifold. Also, such ideal SM control is theoretically achievable only with the absolute compliance of the three fundamental conditions of hitting, existence and stability, and the condition that the system operates at an infinite switching frequency. Thus, what is derived is an idealized controlled system, for which no external disturbance or system's uncertainty can affect the *ideal control* performance of having a precise tracking, zero-regulation error (infinite DC gain), and very fast dynamic response. In a certain sense, the SM control is an ideal (optimal) type of control for variable structure systems.

1.3.2 Practical Limitations and Chattering

What has been considered so far is based on the ideal assumption that infinite switching frequency is possible and that all relevant components related to the control are perfect. However, in real-world applications, this is not possible. In practice, due to imperfections of switching devices like time delay, response time constant, presence of dead zone, hysteresis effect, and saturation of device switching frequency, the actual behavior of the sliding motion deviates slightly from that expected for the ideal condition. In addition, a kind of high-frequency oscillation may occur in the control process which is reflected in the actual behavior of the trajectory. This phenomenon is known as *chattering* [102].

Figure 1.2 shows the behavior of the trajectory of the controlled system in the sliding phase under non-ideal condition. Note that the non-ideality of switching does not affect the reaching phase of the SM operation, and therefore, under both ideal and non-ideal conditions, the reaching phase can be represented by Fig. 1.1(a). However, for the sliding phase under the non-ideal condition (see Fig. 1.2(a)), trajectory S does not move exactly on the sliding manifold, but instead oscillates within its vicinity at a high frequency while concurrently converging toward O. Additionally, unlike in the ideal condition where trajectory S stops precisely at O upon arrival, in the non-ideal condition, trajectory S will be trapped in a periodically-oscillating state at a point near O. The consequence to this is that while in the ideal condition, the controlled system is maintained in a static position at the desired point O (i.e., no error) during steady state, but in the non-ideal condition, the controlled system is oscillating at a static point slightly away from the desired point, resulting in the presence of steady-state error.

1.3.3 Constant Dynamics

An interesting feature of the SM control is the possession of constant dynamics during the sliding phase. As illustrated in Fig. 1.1(b), when the trajectory touches the manifold and the system enters into the sliding phase, the movement of the trajectory is confined along the sliding manifold, which means

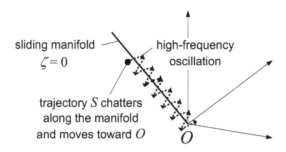

(a) Sliding phase under non-ideal condition

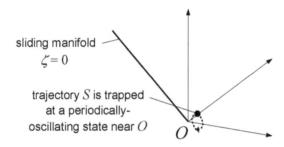

(b) Steady-state trajectory under non-ideal condition

FIGURE 1.2
Graphical representations of SM control process under non-ideal condition: (a) Sliding phase—illustrating trajectory S chattering within the vicinity of sliding manifold but still converging toward the origin O; and (b) steady state—trajectory S being trapped in a periodically oscillating state near the origin O.

that the motion equation of the trajectory is basically $S = \zeta = 0$. This is true regardless of the operating conditions and parameter variations in the system. Therefore, the dynamics of a system under SM operation is constant and is independent of the system parameters or disturbance.

Note that such a property applies only to the sliding phase but not the reaching phase which has a different set of dynamic characteristics for a given operating condition. However, since the time taken to complete the sliding phase is typically much longer than the reaching phase, it is sufficient to consider only the dynamics of the former for the control design. Therefore, it is possible to have a system that can attain a constant dynamical behavior using the SM control.

1.3.4 Quasi-Sliding Mode Control

An ideal SM control operates theoretically at an infinite switching frequency so that the trajectory follows exactly the reference track [102]. This requirement for operating at infinite switching frequency, however, challenges the feasibility of applying SM controllers in many systems. This is because extreme high-speed switching may result in excessive switching losses and may be a source of noise that interferes with the operation of the system. Hence, for SM control to be applicable to practical systems, the switching frequency of the control implementation must be confined within a practical range. Nevertheless, this practical requirement on the SM controller's switching frequency essentially makes the control a quasi-sliding mode (QSM) or pseudo-sliding mode (PSM) control, which operates as an approximation of the ideal SM control. The consequence of this approximation is the degradation of the system's robustness and the deterioration of the regulation properties. Clearly, the approximation of QSM to the ideal SM controller becomes better as the switching frequency increases.

1.4 Mathematical Formulation

Consider a nonlinear time-dependent switching system defined by the following equation:

$$\dot{\mathbf{x}}(t) = g(\mathbf{x}(t)) + \varphi(\mathbf{x}(t)) \cdot u(t), \qquad (1.1)$$

where $\mathbf{x}(t)$ is the state-variable vector in an n-dimensional space R^n; $g(.)$ and $\varphi(.)$ are smooth vector fields in the same space; and $u(t)$ is the discontinuous control action expressed as

$$u(t) = \begin{cases} U^+ & \text{if } S(\mathbf{x},t) > 0 \\ U^- & \text{if } S(\mathbf{x},t) < 0 \end{cases} ; \qquad (1.2)$$

where U^+ and U^- are either scalar values or scalar functions of $\mathbf{x}(t)$; and $S(\mathbf{x},t)$ is the instantaneous feedback-tracking trajectory of the system and is a predetermined function of the state variables. Typically, for ease of design and implementation, $S(\mathbf{x},t)$ is chosen as a linear combination of the weighted values of the state variables, and is given as

$$S(\mathbf{x},t) = \sum_{i=1}^{m} \alpha_i x_i(t), \qquad (1.3)$$

where α_i for $i = 1$ to m denotes the set of the control parameters known as sliding coefficients and $x_i(t) \in \mathbf{x}(t)$. A system with this description is said to exhibit SM property when all the required conditions, namely hitting, existence, and stability conditions, are met.

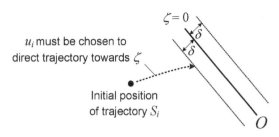

FIGURE 1.3
Trajectory S converging to the sliding manifold in SM control process when the hitting condition is fulfilled.

1.4.1   Hitting Condition

The objective of the hitting condition is to ensure that regardless of the location of the initial condition, the corresponding control decision will direct the trajectory of the system to approach and eventually reach, within a vicinity δ, the sliding manifold (see Fig. 1.3). Assume that the system is at an initial state with vector $\mathbf{x}_i = \mathbf{x}(t = 0)$ and a trajectory $S_i = S(t = 0)$ which is located at a distance away from the sliding manifold $\zeta = 0$. The necessary and sufficient condition for the system to satisfy the hitting condition is that the resulting control $u_i = u(t > 0)$ produces a state-variable vector $\mathbf{x}(t > 0)$ and consequently a controlled trajectory $S(t > 0)$, which satisfies the following inequality:

$$S\frac{dS}{dt} < 0 \quad \text{(for } t > 0 \text{ and that } |S| \geq \delta\text{).} \tag{1.4}$$

The inequality (1.4) is a partial result of the *Lyapunov second theorem on stability* [21, 68, 85, 101, 102], of which the *Lyapunov function candidate* is

$$V(S) = \frac{1}{2}S^2. \tag{1.5}$$

The compliance of (1.4) signifies that at a position not within the vicinity of the sliding manifold, the state trajectory S is continuously being attracted and is always converging to the sliding manifold $\zeta = 0$ for $t > 0$, and that the choice of $u_i = u(t > 0)$ is supporting this attraction. Hence, one important and fundamental aspect of designing the SM control is to first determine, for a desired set of control parameters (sliding coefficients), the suitable discontinuous control action for the system, as described in (1.2). In other words, the design of U^+ and U^- would have to ensure that the hitting condition be always satisfied for a given system.

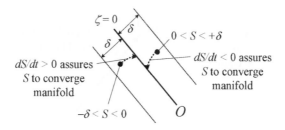

FIGURE 1.4

Trajectory S at a location within the vicinity of sliding manifold $0 < |S| < \delta$ converging to the sliding manifold in SM control process when the existence condition is fulfilled.

1.4.2 Existence Condition

Having the system already designed to fulfil the hitting condition, it is then necessary to check if the system also complies with the existence condition which ensures that once the trajectory is at locations within the vicinity of the sliding manifold such that $0 < |S| < \delta$, it is still always directed toward the sliding manifold, as illustrated in Fig. 1.4.

The existence condition of the SM operation can be determined by inspecting only the local reachability condition of $S\frac{dS}{dt} < 0$ [21], such that in the domain of $0 < |S| < \delta$, the condition

$$\lim_{S \to 0} S \cdot \frac{dS}{dt} < 0 \tag{1.6}$$

must be satisfied. This can be expressed as

$$\lim_{S \to 0^+} \frac{dS}{dt} < 0 \quad \text{and} \quad \lim_{S \to 0^-} \frac{dS}{dt} > 0. \tag{1.7}$$

In a physical sense, the existence condition can be understood as being a requirement for the controlled trajectory and its time derivative to have opposite signs in the vicinity of a discontinuous surface [102].

1.4.3 Stability Condition

In addition to the existence condition, the control action and sliding coefficients must be designed to comply with the stability condition. This is to ensure that in the event of operating in the sliding phase, the desired sliding manifold will always direct the trajectory toward a stable equilibrium point. Failure to achieve this will lead to an SM system which is unstable. Figure 1.5(a) shows the trajectory of a system under SM operation stabilizing at the desired point of equilibrium O when stability condition is fulfilled, and Fig. 1.5(b) shows the same trajectory moving pass O when stability condition is not fulfilled.

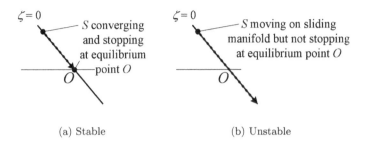

(a) Stable   (b) Unstable

FIGURE 1.5
Trajectory S moving on the sliding manifold and (a) converging to the desired
point of equilibrium O when stability condition is fulfilled, and (b) passing
through and not stopping at the desired point of equilibrium O when stability
condition is not fulfilled.

In general, the stability of a system is obtained by ensuring that the *eigen-values* of the *Jacobian* of the system at the steady-state region have negative
real parts. In the following, we will consider how the stability condition of the
SM operation can be found for a system with a linear sliding manifold and
for a system with a nonlinear sliding manifold.

1.4.4 System with Linear Sliding Manifold

For an SM-controlled system whose trajectory is made up of the state variables, their time derivatives/subderivatives, and/or time integrals/iterated-
integrals (so-called in the phase canonical form), its equivalent sliding manifold
is basically a linear motion equation, which can be obtained by substituting
$S = 0$ into (1.3), i.e.,

$$\alpha_1 x_1(t) + \alpha_2 x_2(t) + \ldots + \alpha_m x_m(t) = 0. \qquad (1.8)$$

Since the state variables $x_{n=1,2,3,\ldots,m}$ are in a phase canonical form such that
$x_{n+1}(t) = \frac{dx_n}{dt}(t)$, we have

$$\alpha_1 X_1(s) + \alpha_2 s X_1(s) + \ldots + \alpha_m s^{m-1} X_1(s) = 0 \qquad (1.9)$$

where $X_1(s)$ is the Laplace transform of $x_1(t)$, and (1.9) can be arranged and
simplified as

$$\alpha_m s^{m-1} + \ldots + \alpha_3 s^2 + \alpha_2 s + \alpha_1 = 0. \qquad (1.10)$$

By applying the *Routh-Hurwitz stability criterion* to this $m - 1^{\text{th}}$ order linear
polynomial, the condition for stability which requires that all roots of this
characteristic equation have negative real parts can be obtained. For example,
for a second-order polynomial, the stability condition would be $\alpha_1 > 0$ and
$\alpha_2 > \alpha_3 > 0$.

1.4.5 System with Nonlinear Sliding Manifold

For an SM-controlled system having a nonlinear motion equation during SM operation, a different approach based on the *equivalent control method* is adopted to derive the stability condition [48, 49]. This approach involves first deriving the *ideal sliding dynamics* of the system, and then performing a stability analysis on its *equilibrium point*.

1.4.5.1 Ideal Sliding Dynamics

The replacement of the discontinuous control action $u(t)$ by a continuous control action $u_{eq}(t)$ into (1.1) converts the switching SM system into an average continuous SM system given as

$$\dot{\mathbf{x}}(t) = g(\mathbf{x}(t)) + \varphi(\mathbf{x}(t)) \cdot u_{eq}(t). \tag{1.11}$$

The control action $u_{eq}(t)$, which is the *equivalent control* derived from the so-called *equivalent control method* (to be discussed in the following section), is a solution of $\frac{dS}{dt} = 0$. From (1.3), it is clear that $u_{eq}(t)$ is a function $f(.)$ of the dynamics of the state variables and the sliding coefficients, and it can be given as

$$u_{eq}(t) = f\left(\sum_{i=1}^{m} \alpha_i \dot{x}_i(t) = 0\right). \tag{1.12}$$

Then, substitution of (1.12) into (1.11) gives

$$\dot{\mathbf{x}}(t) = g(\mathbf{x}(t)) + \varphi(\mathbf{x}(t)) \cdot f\left(\sum_{i=1}^{m} \alpha_i \dot{x}_i(t) = 0\right) \tag{1.13}$$

which represents the ideal sliding dynamics of the system during SM operation and is independent of the control signal.

1.4.5.2 Equilibrium Point

Assume there exists a stable equilibrium point on the sliding manifold on which the ideal sliding dynamics eventually settled. Then, the state equations in (1.13) can be solved to give the steady-state operating point $(x_{1(ss)}, x_{2(ss)}, ..., x_{m(ss)})$ during SM operation by putting $\dot{\mathbf{x}}(t) = 0$.

1.4.5.3 Linearization of Ideal Sliding Dynamics

For a system with nonlinear SM motion equation, the ideal sliding dynamics will also inherit the nonlinearity. To determine the stability of the equilibrium point, it is necessary to first obtain the linear description of the ideal sliding dynamics around the equilibrium point.

The linearization of the ideal sliding dynamics around the equilibrium point $(x_{1(ss)}, x_{2(ss)}, ..., x_{m(ss)})$ by using *perturbation theory* transforms equation (1.13) into

$$\dot{\tilde{\mathbf{x}}}(t) = h(\tilde{\mathbf{x}}(t)) + \psi\left(\tilde{\mathbf{x}}(t), \sum_{i=1}^{m} \tilde{x}_i(t)\right) \qquad (1.14)$$

where $\dot{\tilde{\mathbf{x}}}(t)$ represents the linearized small-signal ideal sliding dynamics around the steady-state operating point, and $h(\tilde{\mathbf{x}}(t))$ and $\psi\left(\tilde{\mathbf{x}}(t), \sum \tilde{x}_i(t)\right)$ are the small-signal ac equivalent components of $g(\mathbf{x}(t))$ and $\varphi(\mathbf{x}(t)) \cdot f\left(\sum \alpha_i \dot{x}_i(t) = 0\right)$ in (1.13), respectively.

Arranging (1.14) in the matrix form, i.e.,

$$\dot{\tilde{\mathbf{x}}}(t) = \mathbf{A}\tilde{\mathbf{x}}(t) \qquad (1.15)$$

where $\mathbf{A}$ is the Jacobian square matrix of $\tilde{\mathbf{x}}(t)$, the characteristic equation of the linearized sliding dynamics is given as

$$\det(\mathbf{A} - \lambda I) = 0 \qquad (1.16)$$

where $\det(.)$ represents the determinant function, I represents the identity matrix, and λ represents the eigenvalue of the system. The application of the *Routh-Hurwitz stability criterion* to the characteristic equation (1.16) will result in a set of conditions that will ensure the stability of the ideal sliding dynamics and hence confirm the presence of a stable equilibrium point on the sliding manifold.

1.4.5.4 Remarks

Note that the aforedescribed approach for evaluating the stability of the nonlinear sliding manifold is applicable to all types of sliding manifolds including the linear sliding manifold, and the same result given in Section 1.4.4 will be valid for all types of sliding manifolds.

Moreover, the stability of the system under SM operation can also be inherently accomplished through the design of the sliding coefficients to meet the desired dynamical property [1]. The procedure essentially makes use of the invariance property of SM operation. Specifically, as the state trajectory S moves along the sliding manifold toward a stable point, the motion equation $S = 0$, which describes the dynamical property of the system can be obtained. Thus, with a proper selection of the sliding coefficients, the system will respond desirably and the stability condition is inherently met.

Finally, it must be emphasized that in any design of the SM control, both the stability condition and the existence condition are the basis for the design of the sliding coefficients.

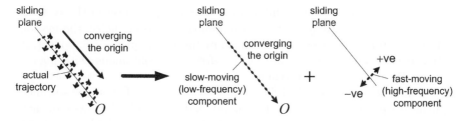

FIGURE 1.6
Graphical representation of high- and low-frequency components of the trajectory in the sliding phase.

1.5 Equivalent Control

The ideal SM control operation is supposed to operate the system at an infinite switching frequency such that the trajectory moves precisely on the sliding manifold. However, as discussed before, practical limitations of devices and components in the system will alter the actual behavior of the sliding motion and induce a low-amplitude high-frequency oscillation (chattering) within the vicinity of the sliding surface while moving toward the origin (see Fig. 1.2(a)). It is possible to identify two components in the trajectory, namely a "fast-moving" (high-frequency) component and a "slow-moving" (low-frequency) component.

Figure 1.6 shows a graphical representation of the trajectory. It can be seen from the diagram that the high-frequency component oscillates between the +ve and −ve directions, whereas the low-frequency component moves along the sliding plane. Since the movement of the trajectory is a consequence of the switching action $u(t)$, it is therefore possible to relate the corresponding low-frequency and high-frequency components of the trajectory to a low-frequency continuous switching action $u_{\text{low}}(t)$, which satisfies

$$U^- < u_{\text{low}}(t) < U^+ \qquad (1.17)$$

and a high-frequency discontinuous switching action defined by

$$u_{\text{high}}(t) = \begin{cases} U^+ - u_{\text{low}}(t) & \text{if } S(\mathbf{x}, t) > 0 \\ U^- - u_{\text{low}}(t) & \text{if } S(\mathbf{x}, t) < 0 \end{cases} \quad ; \qquad (1.18)$$

and the trajectory is thus given by

$$u(t) = u_{\text{low}}(t) + u_{\text{high}}(t). \qquad (1.19)$$

Clearly, the switching action of $u_{\text{high}}(t)$ produces the high-frequency trajectory component, and the switching action of $u_{\text{low}}(t)$ produces the low-frequency trajectory component. By ignoring the high-frequency component, which can often be practically filtered out by a low-pass filter of the system[1] or otherwise will simply appear as an unwanted chattering noise, the motion of the trajectory is then solely determined by the low-frequency component. It is thus reasonable to consider only the low-frequency continuous switching action $u_{\text{low}}(t)$ as the desired switching action that will produce a trajectory that is *nearly equivalent* to an ideal SM-controlled trajectory. This is the so-called *equivalent control* of the system, i.e., $u_{\text{eq}}(t)$, and is actually the low-frequency continuous switching action $u_{\text{low}}(t)$ described above. The method of obtaining the equivalent control is given as follows.

Consider a system $\dot{x}(t) = f(x(t), u(t))$, where $u(t) = (U^+, U^-)$. Under an ideal SM operation, the trajectory $S(t)$ is always moving along the sliding manifold, i.e., $S(t) = 0$, and in the absence of any high-frequency oscillation, it is also true that $\frac{dS(t)}{dt} = \dot{S}(t) = 0$. In SM control theory, this is also known as the *invariance conditions*.

Next, suppose that the equivalent control produces a trajectory resembling an ideal SM operation. Under such condition, the system equation $\dot{S}(t) = G \cdot f(x(t), u(t))$, where $G = \frac{\partial S(t)}{\partial x(t)}$, can be rewritten as $\dot{S}(t) = G \cdot f(x(t), u_{\text{eq}}(t))$. Then, the solution of the equivalent control $u_{\text{eq}}(t)$ can be obtained by solving $G \cdot f(x(t), u_{\text{eq}}(t)) = 0$ and can be expressed as $u_{\text{eq}}(t) = H(G \cdot f(x(t), u_{\text{eq}}(t)) = 0)$.

Finally, if $u_{\text{eq}}(t)$ is substituted back into the original system, we get

$$\dot{x}(t) = f(x(t), u_{\text{eq}}(t)), \tag{1.20}$$

which is the motion equation of the system in SM operation. This method of deriving the equivalent control signal u_{eq}, along with the formulation of the motion equation as shown in (1.20), is known as the *equivalent control method* [102].

1.6 Types of Implementation

1.6.1 Relay and Signum Functions

The conventional method of implementing the SM control is based directly on the control law described in (1.2) and (1.3), of which the former is simply a discontinuous function that is easily realized using a switch relay and the latter, which computes the instantaneous trajectory $S(t)$, is realized through analog or digital computation (see Fig. 1.7).

In many applications where the control involves only a positive-or-negative decision, i.e., $U^+ = 1$ and $U^- = -1$, the *signum function* can be used for the

---

[1]Note that in the case of power electronics converters, the high-frequency ac signals are filtered out by the output filter capacitor.

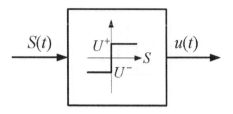

FIGURE 1.7
A relay function in SM control.

relay, i.e.,

$$u(t) = \text{sgn}(S(\mathbf{x}, t)), \tag{1.21}$$

where the signum function sgn(.) is defined as

$$u(t) = \begin{cases} 1 & \text{if } S(\mathbf{x}, t) > 0 \\ 0 & \text{if } S(\mathbf{x}, t) = 0 \\ -1 & \text{if } S(\mathbf{x}, t) < 0 \end{cases} . \tag{1.22}$$

For applications where the control involves only digital logic, (1.22) is replaced by

$$u(t) = \begin{cases} 1 & \text{if } S(\mathbf{x}, t) > 0 \\ 0 & \text{if } S(\mathbf{x}, t) \leq 0 \end{cases} . \tag{1.23}$$

In general, the implementation of the control using this approach is straightforward and simple. However, as discussed earlier, the direct implementation of this control law results in systems that are switched at a very high frequency giving an unwanted chattering effect in the system [90, 102]. This makes it unsuitable for some applications which see this as an undesired high-frequency noise. Therefore, for such systems, it is necessary to restrict the range of the operating frequency, for instance, by using a hysteresis function.

1.6.2 Hysteresis Function

The implementation of SM control through hysteresis function does not require additional computation or auxiliary circuitries, and its implementation is easily accomplished by redefining (1.2) as

$$u(t) = \begin{cases} U^+ & \text{if } S(\mathbf{x}, t) > \Delta \\ U^- & \text{if } S(\mathbf{x}, t) < -\Delta \\ \text{Previous state} & \text{Otherwise} \end{cases} ; \tag{1.24}$$

where Δ is an arbitrarily small value. The introduction of a hysteresis band with the boundary conditions $S = \Delta$ and $S = -\Delta$ provides a form of control

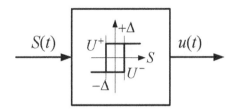

FIGURE 1.8
A hysteresis function in SM control.

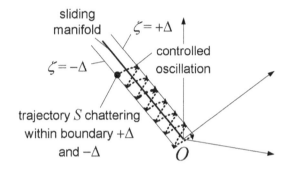

FIGURE 1.9
Trajectory in SM control with hysteresis function.

to the switching frequency of the system, thus solving the practical problem of very high-frequency switching operation since the time period for the interchanging of the states is now delayed (see Fig. 1.8). As a result, the trajectory S of the system will operate precisely in the vicinity of $\pm\,\Delta$ of the sliding manifold with a controlled oscillation as illustrated in Fig. 1.9. The chattering effect now becomes controllable and is a function of Δ.

1.6.3   Equivalent Control Function

Instead of implementing the original SM control based on (1.2) and (1.3), it is also possible to implement the control using a derived equivalent control [78]. As discussed in Section 1.5, the equivalent control function $u_{eq}(t)$ of a system under SM control will be the ideal averaged control function required during the SM phase and is bounded by $U^- < u_{eq}(t) < U^+$. Based on the invariance conditions $S = 0$ and $\dot{S} = 0$, the equivalent control can be derived by simply solving $\dot{S} = 0$. Theoretically, the implementation of the equivalent control function $u_{eq}(t) = H(G \cdot f(x(t), u_{eq}(t)) = 0)$ will result in a system that operates with ideal SM control without high-frequency chattering.

It is interesting to note that the derivation of the equivalent control using only part of the invariance conditions $\dot{S} = 0$ (and not $S = 0$) would be suf-

FIGURE 1.10

Photograph of the semiconductor die of the pulse-width modulation-based SM controller IC.

ficient for implementing the SM control. This is because when the equivalent function evolves on S, there is an interdependency among the state variables of the ideal sliding system (with $S = 0$ and $\dot{S} = 0$) and therefore, one of the equations in the invariance conditions is redundant [78]. In other words, when a system enters into the SM phase, i.e., $S = 0$, the implementation of a control law that ensures $\dot{S} = 0$ will automatically imply $S = 0$, and vice versa.

For a clearer understanding, consider a physical system. A car with a velocity error of $v_e = 0$ ($S = 0$) with a control law that keeps its acceleration error at $a_e = 0$ ($\dot{S} = 0$) will have its velocity error kept indirectly at $v_e = 0$ (S = 0) since there is no error in acceleration. This is a clear reflection on the interdependency relationship of the two variables, and it is obvious that there is no necessity to make a specific control on the velocity error. In essence, a control law implementing $\dot{S} = 0$ will indirectly ensure that $S = 0$, and vice versa. As a final remark, the overall control design should ensure that $S = 0$ in the first place through the compliance of the hitting and existence conditions. This in turn will ensure that the controlled trajectory is always attracted to and will arrive at the sliding manifold, and will be kept within its vicinity.

In the subsequent chapters, we will describe systematically the theory, analysis, design, and implementation of SM control specifically for power electronics applications. Eventually, we wish to illustrate the practicality of using SM control for low-cost power converters applications. At the time of writing, successful realization of fixed-frequency SM controllers in analog integrated-circuit (IC) forms has been achieved for DC–DC converters operating under wide variations of input voltage and load. Figure 1.10 shows the semiconductor die of the first pulse-width modulation-based SM controller IC for DC–DC converters.

2

Overview of Power Converters and Their Control

CONTENTS

2.1 Introduction

The role of power conversion is to facilitate the transfer of power from the source to the load by converting voltages and currents from one magnitude and/or frequency to another. This function of power processing is performed through an analog power circuit known as the power converter (see Fig. 2.1). A controller is required for the management of this power transfer process. The ultimate aim of the entire conversion process is to have as high an efficiency as possible, while achieving as closely as possible the desired conversion and control functions.

DC–DC (pronounced DC-to-DC) converters are power electronic circuits that accept DC input voltages or currents and produce DC output voltages

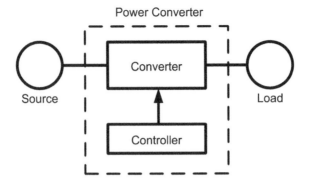

FIGURE 2.1
Power converter.

or currents. Typically, the magnitude of the output is different from that of
the input. In addition, DC–DC converters are sometimes employed to provide
noise isolation, galvanic isolation and power flow regulation. The following
section gives a brief discussion on the various circuit topologies of DC–DC
converters.

2.2   Basic DC–DC Converters

According to Liu *et al.* [39], there are six possible basic configurations of DC–
DC converters, namely, the buck, boost, buck-boost, Ćuk, Sepic, and Zeta
converters. However, since the buck, boost, and buck-boost converters are
the simplest and most commonly used converters for power regulation, and
that Ćuk, Sepic, and Zeta converters can be constructed by combining these
converters, we will limit our discussion to the buck, boost, and buck-boost
converters.

As shown in Fig. 2.2, each of these converters consists of only one active
(bidirectional) power switch S_W, one passive (unidirectional) power switch,
usually a diode D, one inductive storage element L, and one capacitive storage
element C. Here, v_i is the input voltage, v_o is the output voltage, and r_L is
the load, to which the power is being delivered.

The primary function of the buck converter is to step down an input voltage
to a lower output voltage, i.e., $v_o < v_i$. Conversely, the primary function of the
boost converter is to step up an input voltage to a higher output voltage, i.e.,
$v_o > v_i$. The buck-boost converter, as the name suggests, allows both stepping
up and down of the input voltage. In all three converters, the magnitude of
the voltage conversion is directly controlled by the turning on and off of the
switch. An important point to note is the energy transfer property of these

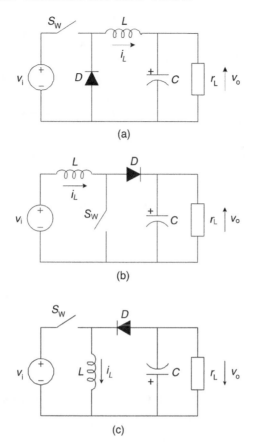

FIGURE 2.2
Basic DC–DC converter topologies: (a) buck converter; (b) boost converter; (c) buck-boost converter.

converters. For the buck converter, the energy is directly transferred to the output and the capacitive storage at the instance when switch S_W is turned on. This happens concurrently with the energizing process of its inductive storage element. However, for the boost and buck-boost converters, only the energizing process takes place when switch S_W is turned on. The process whereby the energy is transferred from the inductive storage to the output and the capacitive storage happens only after S_W is turned off. This indirect energy transfer from the source to the load, via the inductor, results in a phase lag phenomenon commonly understood as the right-half-plane zero (RHPZ) or non-minimum phase response property.

Furthermore, in the presence of two switches, these converters can be visualized as multi-structural systems, in which each structure comprises a linear

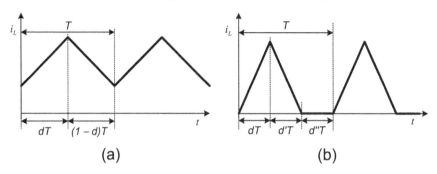

(a)   (b)

FIGURE 2.3
Inductor current waveform during (a) continuous conduction mode operation
and (b) discontinuous conduction mode operation.

circuit configuration. The change of the circuit configuration is basically governed by the setting of the switches. For any given switching condition, there
are two independent first-order differential equations that describe the behavior of the converter.

2.3   Operating Modes of DC–DC Converters

There are two operating modes in the buck, boost, and buck-boost converters,
namely, the continuous conduction mode and the discontinuous conduction
mode. When operating in continuous conduction mode, there exist only two
switching states in the converter, i.e., "S_W is on and D is off" and "S_W is
off and D is on." The converter can remain in this operating mode for as
long as the inductive storage is relatively large. Conversely, if the inductive
storage is relatively small, the operation falls into discontinuous conduction
mode, resulting in three switching states, i.e., "S_W is on and D is off", "S_W
is off and D is on," and "S_W is off and D is off."

The mode of operation can be identified by inspection of the inductor current waveform. Figure 2.3(a) shows the typical inductor current waveform of a
converter operating in continuous conduction mode. Here, the turn-on time of
the active switch S_W is denoted as dT, where d is the duty ratio (also known
as duty cycle) and T is the switching period. As seen from the figure, a major
characteristic of the continuous conduction mode operation is that its inductor current i_L is always greater than zero. In principle, this is only possible
if the inductance is sufficiently large, or the switching period is sufficiently
short. On the other hand, if the inductance is relatively small, or the period
is relatively long, there will be a period of time when the inductor current
falls momentarily to zero, as shown in Fig. 2.3(b). In such a circumstance, the

converter is said to be operating in discontinuous conduction mode. Here, $d'T$ represents the time duration in a switching cycle when diode D is conducting and power switch S_W is opened. Additionally, $d''T$ represents the time duration when both the diode and the switch S_W are opened. This time period can be equivalently expressed as $(1 - d - d')T$.

2.4   Overview of Control

Most DC–DC converters are designed with a closed-loop feedback controller to deliver a regulated output voltage. The main objective is to ensure that the converter operates with a small steady-state output error, fast dynamical response, low overshoot, and low noise susceptibility, while maintaining high efficiency and low noise emission. All these design criteria can be achieved through the proper selection of control strategies, circuit parameters, and components.

Before moving on to the discussion on control methodologies, it is important to understand the various factors that affect the control performance.

2.5   Factors Influencing Control Performances

This section discusses how the switching frequency, the energy storage elements of the converter, the gain parameters, and the type of the controller would affect the control performance of a converter.

2.5.1   Switching Frequency

Theoretically, *ideal control* is achievable only if the switching frequency is infinitely high. The term ideal control, in our context, would mean perfect DC steady-state regulation (i.e., zero error and noise susceptibility), infinitely fast dynamic response, and no overshoot.

However, since all practical systems are subjected to time delay and slew rate limitation of circuit components, an infinitely high switching frequency operation is unattainable. The magnitude of the switching frequency is therefore limited by the bandwidths of the circuit components, i.e., the controller, power switch, diode, etc. Moreover, it is also well known that the amount of power loss in a converter increases as the switching frequency increases. This is true even if *soft-switching techniques* are applied to reduce the switching losses of the power switches and/or diodes. It should be noted that *eddy current* and *hysteresis current* losses of the magnetic components and *ac re-*

sistive power losses of the circuit due to skin effect will increase as switching frequency increases. Of course, all these losses may be relatively lower than the losses generated by the switches. Nevertheless, it should be considered when determining the switching frequency of a converter since they will become significant when the frequency is very high. Furthermore, the emission of high-frequency noise may cause undesirable radio and electromagnetic interferences. All these factors will limit the practicality of very fast switching frequency to achieve an ideal regulation.

Fortunately, in most applications, an ideal power converter is often not needed. What is actually required is a power converter that meets some line/load regulation criteria, ac ripple criteria, dynamic response criteria, and size, weight, and efficiency criteria. Therefore, the selection of the switching frequency should always be a tradeoff between achieving the desired control performance to meet consumer specifications and minimizing power losses, power density, and electromagnetic interference. Nevertheless, it is true that the control performance of a converter can always be improved with a higher switching frequency.

2.5.2 Energy Storage Elements

The size of the inductive and capacitive energy storage elements, i.e., L and C can also affect the control performance of a DC–DC converter. This can be understood by examining the operating mechanism of these elements. Since the rate of storing/releasing electrical energy, i.e., $\tau = r_L C$ (capacitor) and $\tau = \frac{L}{r_L}$ (inductor), is directly affected by the size of the energy storage elements, the ability to respond to the load changes is therefore also affected by the size of the energy storages. Specifically, for a fixed-frequency operation, the dynamic response of the converter will generally be faster with a smaller value of L or C, since a smaller energy storage element requires a shorter time to store and release energy. On the other hand, a smaller value of L will lead to a higher-ripple inductor current and a smaller output voltage undershoot or overshoot during a step increment or step decrement in the load current. Yet, a larger value of C will give a smaller output voltage undershoot during a step increment in the load current, and a smaller output voltage overshoot during a step decrement in the load current. Thus, the size of the storage elements together with the control design determine the dynamical behavior of the controlled converter. In fact, there have been some proposals about using a variable inductor to alter the dynamical response of power converters.[1]

2.5.3 Control Gains

It is well known that in any control methodology, the choice of the control gain parameters play an important role in determining the response of the

---

[1] US patent 6 188 209, issued on February 2001.

system. It is therefore necessary to understand how these control gains can ultimately affect the control performance of the system.

First of all, regardless of the control methodology, there are only two types of parameters in a feedback control scheme, namely, the *controlled variables* and the *control gains*. The controlled variables are basically the variables to be controlled. In the case of a DC–DC converter, these are usually the output voltage error and/or the inductor current error. The main objective, from the perspective of the controller, is to ensure the nullification of the output voltage error and/or the inductor current error for any disturbance or change, in a stable manner within a shortest time. This is achieved through the manipulation of the controlled variables, by adopting their direct, integral, and/or derivative forms as *manipulated variables* in the control computation [65]. The purpose of the control gains is to act as multiplication factors to amplify these manipulated variables, so that the effects of varying these manipulated variables can be intensified. This provides the shaping of the controlled behavior such that the desired response can be achieved.

Assume that the output voltage error V_e is the only controlled variable. Under classical control theory for continuous systems, increasing the control gain K_p of the controlled variable, i.e., the output voltage error V_e, improves the transient response and reduces the steady-state error. However, a high K_p also reduces the stability of the system. Hence, it is a choice between better response and poorer stability. Next, increasing the control gain K_i of the time integral of the controlled variable, i.e., $\int V_e dt$, speeds up the elimination of the steady-state error, but also reduces the stability of the system. Finally, increasing the control gain K_d of the time derivative of the controlled variable, i.e., $\frac{dV_e}{dt}$, reduces the overshoots and oscillation of the control performance. K_d does not affect the steady-state performance of the system, but it is sensitive to noise. The above description is valid for both large-signal nonlinear controlled systems and small-signal linear controlled systems.

Moreover, it is important to note that while the above-described control behavior holds in general for switching systems like the DC–DC converters, it it not exact. Contrary to a continuous system, the incorporation of an integral control term into the controller of a DC–DC converter can only *reduce*, but not *eliminate*, the DC steady-state error. There exists a finite DC steady-state error at the output of the converter. The failure of the integral control in providing an absolute zero DC steady-state error condition (see Fig. 2.4) in a switching system is attributed to the finite switching operation, which fails to generate an instantaneous error correction. Consequently, the inevitable DC steady-state error makes the line and load regulations imperfect. As previously mentioned, an ideal control can only be achieved at infinite switching frequency. This is particularly true in this limiting case since the limit cycle operation of finite switching will cease at infinite switching frequency, ending in a theoretically zero steady-state error and perfect regulation.

Additionally, the discussion above is from a classical proportional-integral-derivative (PID) control viewpoint. In practice, it is common that the control

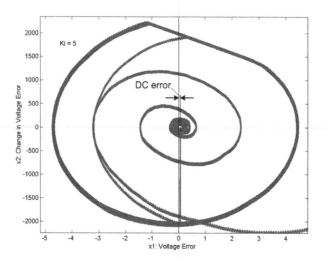

FIGURE 2.4
Phase portrait of the output voltage error and its derivative of a buck converter
with integral voltage control in its steady-state operation.

gains of power converters are designed in the frequency domain through the
selection of a suitable composite transfer function $K\frac{(s+z_1)(s+z_2)\cdots(s+z_n)}{(s+p_1)(s+p_2)\cdots(s+p_m)}$ made
up of a DC gain K; poles $\frac{1}{s+p_1}$, $\frac{1}{s+p_2}$, $\cdots$, and $\frac{1}{s+p_m}$; and/or zeroes $s + z_1$,
$s + z_2$, $\cdots$, and $s + z_n$ such that the overall system has an adequately high
DC gain and is within the desired phase and gain margins. Nevertheless, by
converting the composite transfer function back into its time domain representation of the proportional, integral, and derivative forms, it is possible to
understand how the poles and zeroes can affect the overall behavior of the
controlled system from the classical control viewpoint.

2.6   Common Control Techniques

This section presents the various techniques used in the control of DC–DC
converters, as well as the approaches for designing the controllers.

2.6.1   Hysteretic Controllers

Prior to the introduction of fixed-frequency pulse-width modulation (PWM)
integrated-circuits (IC) controllers, the most commonly used control technique

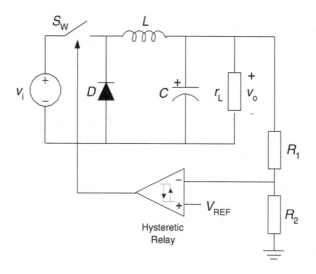

(a) Hysteretic voltage-mode control

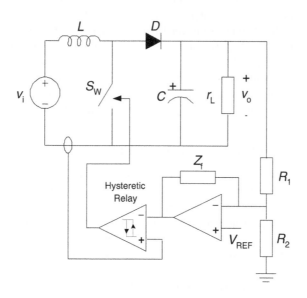

(b) Hysteretic current-mode control

FIGURE 2.5
Simplified hysteretic controllers: (a) hysteretic voltage controlled buck converter and (b) hysteretic current controlled boost converter.

was perhaps the hysteretic controller. There are two types of hysteretic controllers, namely, the *hysteretic voltage controller* and the *hysteretic current controller*. Probably the simplest method of regulating a DC–DC converter, the hysteretic voltage controller, as shown in Fig. 2.5(a), comprises only a hysteretic relay, which compares the actual output voltage with the desired output voltage. If the actual output voltage becomes too low, power switch S_W turns on; and if the actual output voltage becomes too high, power switch S_W turns off. However, this solution is only effective for buck-type converters where there is no phase lag in the energy transfer. On the other hand, for converters with RHPZ, i.e., the boost and buck-boost types of converters, the hysteretic voltage controller becomes inadequate for providing good regulation. Recall that there is a phase lag between the control action and the response of the output voltage in converters with RPHZ, while there is no phase lag between the control action and the response of the inductor current [71]. Therefore, for converters with RHPZ, the hysteretic current controller[2], as shown in Fig. 2.5(b), is required. Here, the inductor current is compared with the compensated voltage error. If the actual inductor current becomes too low, power switch S_W turns on; and if the actual inductor current becomes too high, power switch S_W turns off. As a result, very tight regulation of the inductor current, hence the output voltage, is possible.

However, neither the turn-on nor turn-off times of the hysteretic relay are fixed. The frequency of operation is varying and is affected by the converter parameters, i.e., v_i, r_L, L, and C, as well as the hysteresis band. Hence, although hysteretic controlled converters are simple to implement and respond rapidly to sudden load changes, they have unpredictable noise spectrum, making "electromagnetic interference" control more difficult.

2.6.2   Pulse-Width Modulation Controllers

The fixed-frequency PWM control is by far the most popular control technique used for the regulation of DC–DC converters. The two reasons for this are: (1) the availability of low-cost highly sophisticated fixed-frequency PWM IC controllers; and (2) the growing need to minimize the spurious emissions of switching converters in increasingly sensitive computational and communicative environments. It is well known that switching converters are severe noise generators. The task of containing noise is made easier by fixed-frequency operation. With proper filtering, grounding, bonding, and shielding, switching converters can be successfully used in "electromagnetic interference" sensitive applications [58].

Similar to hysteretic controllers, fixed-frequency PWM controllers assume two specific forms, namely voltage feedback control and current-programmed control, also known as *voltage-mode control* and *current-mode control*, respectively.

---

[2]US Patent 4 456 872, issued in January 1984.

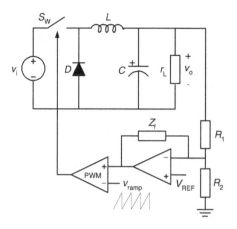

(a) Voltage-mode control

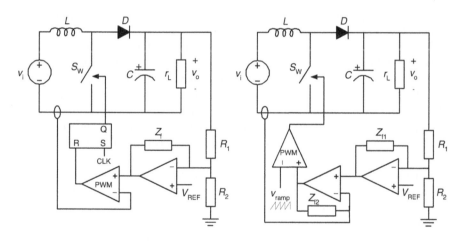

(b) Peak current-mode control   (c) Average current-mode control

FIGURE 2.6
Simplified fixed-frequency PWM controllers for DC–DC converters: (a) voltage-mode controlled buck converter; (b) peak current-mode controlled boost converter; and (c) average current-mode controlled boost converter.

The voltage-mode control is a single-loop control where the output voltage is regulated by closing a feedback loop between the output voltage and the duty-ratio signal (see Fig. 2.6(a)). The output voltage is compared with a constant reference signal V_{ref} to produce an error signal, which is then passed through the compensation network Z_f to generate a control signal. The pulse-width modulator then compares this compensated control signal with an externally generated ramp signal v_{ramp} to generate the desired control signal for driving the power switch. Typically, an input voltage feedforward scheme is required to increase the immunity of the converter output voltage against disturbance in the input voltage. This is accomplished by making the peak value of the ramp signal proportional to the input voltage. Such an input voltage feedforward scheme is not needed in the case of current-mode control, which inherently absorbs the input voltage function into its control consideration.

The current-mode control is a two-loop control system which uses an inner current loop (i.e., inductor current or power switch current) in addition to the voltage feedback loop. In a certain sense, this introduces a nonlinear state feedback term to the controller unlike the PWM voltage-mode controller which is a fully linear controller. As mentioned in the previous section, this kind of control is mainly applied to boost and buck-boost converters which suffer from an undesirable non-minimum phase response. The main advantage of current-mode control is that it increases the overall system's stability (phase) margin, and hence simplifies the design of the outer voltage feedback loop. There are basically two types of fixed-frequency PWM current-mode control in practice, namely, the *peak current-mode control* and the *average current-mode control*.

For the peak current-mode control, the aim is to force the peak inductor current to follow a reference signal which is derived from the output voltage feedback loop (see Fig. 2.6(b)). The idea is to turn on the power switch S_W periodically as commanded by a fixed-frequency clock signal, and to turn it off when the peak of the instantaneous current reaches the desired reference level. One main disadvantage of this method is that it is extremely susceptible to noise, since the current ramp is usually quite small compared to the reference signal. A second disadvantage of the peak current-mode control method is its inherent instability property at duty ratio exceeding 0.5, which results in sub-harmonic oscillation. Typically, a compensating ramp (with slope equal to the inductor current downslope) is required at the comparator input to eliminate this instability. The third disadvantage is that it has a non-ideal loop response because of the use of the peak, instead of the average inductor current sensing.

For the average current-mode control, the Z_{f1} compensated output voltage feedback, which forms the desired current programming level, is first compared to the inductor current to generate the current error (see Fig. 2.6(c)). The pulse-width modulator then compares the Z_{f2} compensated current error to the externally generated ramp signal v_{ramp} to give the desired control signal. The main advantages of this method over the peak current mode method are that it is stable at a duty ratio exceeding 0.5, provides good tracking of average current, and has excellent noise immunity property. However, since there are two compensation networks Z_{f1} and Z_{f1} in series, the analysis and optimal design of these networks are non-trivial. This is a major obstacle for adopting the average current-mode control.

2.6.3 Design Approaches

There are several approaches for designing the compensation networks of the feedback controllers. The most commonly used method is the *pole placement method*, which employs the *Bode plots* and is based on shaping the loop transfer function so as to achieve a desired crossover frequency and phase margin. The first step is to plot the open-loop frequency response of the converter plant using a linearized small-signal model [57]. Next, based on both the ideal open-loop response Bode plots and the converter response Bode plots, an appropriate type of controller is identified. The positions of the poles and zeroes of the compensation network of the chosen controller are then located. This completes the compensation-network design using the pole placement method.

A second approach of designing the compensation network is through the computation of the controller parameters targeted at achieving an optimum closed-loop transfer function for the system, which includes both the small-signal model of the controller and the converter. The objective, as in the pole placement approach, is to design a system that has a suitably fast and stable response.

As easily understood, the reason for the popularity of using small-signal models in design is that the models are linear and can be easily applied in practical design problems. Of course, this would require the additional stages of having the nonlinear model of the converter first averaged, and then linearized about a desired operating point. But overall, it would be a much simpler process than having to design the controller based on the nonlinear models of the converter. Moreover, since the structure of the compensation network is linear, it would be meaningless to design the controller using the nonlinear approach, considering that only one optimal response is possible for a desired operating condition.

2.6.4 Problems of Small-Signal Models and Compensation

Although small-signal models are very convenient for analysis, they have inherent inadequacy. Due to the small-signal assumptions, the converter's behavior cannot be adequately captured when there is a large transient. This can be understood from the criterion for linearity approximation, that is, the time variations, or ac excursions, of the pertinent variables about their respective DC operating points must be sufficiently small. In the case of large-signal operations, the accuracy is lost. Worse still, the small-signal model fails to reveal any stability information of the converter over the entire operating region. In the work by Hamill *et al.* [32], it is shown that even though stable operation is concluded from the small-signal analysis, the system can actually be unstable.

Such drawbacks have motivated power electronics researchers to work along two directions: the search for more accurate linear models or simplified nonlinear models [22, 37, 40, 73, 97, 99, 100, 106, 107] for the controller design, and the investigation into the possible application of different types of nonlinear controllers for DC–DC converters to improve their controllability and performances for a wide operating range.

2.7 Control Methodologies in Research

The failure of conventional linear and partial nonlinear control schemes, namely, PWM voltage and PWM current-mode control, to operate satisfactorily in large-signal operating conditions, is the main motivation for investigating alternative methods of controlling DC–DC converters. This section gives an overview of the various control methodologies currently being considered and have been found to have potential applications in DC–DC converters that require very high performance in dynamical response.

2.7.1 Adaptive Control

An adaptive control scheme is any control scheme that can vary its control parameters and/or control equations according to the changes of the operating conditions. Under such a configuration, it is possible to realize a controller that has an optimal performance for all operating conditions [98]. However, the implementation of this kind of adaptive control is not easy. Not only does it require the sensing of the instantaneous operating condition, but it also requires an effective mathematical formulation of the optimal equations required for generating the optimal operating condition. Therefore, the realization of an adaptive controller demands a highly complex form of analog circuitry, or the use of a digital controller. This explains the unpopularity of the method in practical applications.

2.7.2 Fuzzy Logic Control

The fuzzy logic control is a type of heuristic-reasoning based expert-knowledge automatic control method. It does not require a precise mathematical model of the system or complex computations. The control design is simple since it relies on the designer's understanding of the system's behavior and is based on qualitative linguistic control rules. This extends the control capability to include operating conditions where linear control techniques fail, i.e., large-signal dynamics and parameter variations, and has therefore motivated the investigation of implementing fuzzy logic controllers for DC–DC converters [54, 87]. However, it was found that the main problem with the application of fuzzy logic controllers in DC–DC converters is that its implementation is not practical. It typically involves some digital processes, and hence requires the use of digital signal processor [87] or EEPROM [54], for its implementation.

Even though analog means of implementing the fuzzy logic controller has recently been proposed, the cost of such controllers may still be too high for low to medium power DC–DC converters. Moreover, a second major drawback of the fuzzy logic controller is that it cannot provide, in general, better small-signal response than conventional linear controllers.

2.7.3 Artificial Neural Network Control

An artificial neural network (ANN) is an information processing paradigm that is inspired by the way biological nervous systems process information. It is composed of a large number of highly interconnected processing elements working in unison to solve specific problems. Each of these processing elements, known as the neuron, has a number of internal parameters called weights. Changing the weights of a neuron alters not only the behavior of the neuron, but the behavior of the entire network. Hence, by means of a training process, the weight of each neuron is self-tuned to achieve a desired input–output relationship. Therefore, the main advantage of applying the ANN in the control of DC–DC converters lies in its intrinsic capability to learn and reproduce highly nonlinear transfer function that enables the control to be done effectively for large-signal conditions [14, 38]. However, the main disadvantage of ANN that obscures its practical application in DC–DC converters is that its implementation is non-trivial. Moreover, it requires time-consuming off-line training before it can be used in the actual application.

2.7.4 One-Cycle Control

The idea of *One-Cycle Control*[3] was introduced by Smedley *et al.* [86]. This is basically a nonlinear large-signal PWM method that is designed to control the duty ratio of the switch in real time, such that in each cycle the average of the chopped waveform at the switch output is exactly equal to the control reference. The main advantage of this method is that it is straightforward and the implementation circuits are relatively simple. However, a main drawback of the method is that due the non-ideality of the switches, transistors, and diodes that makes the integration process of the controller non-instantaneous, it is difficult to achieve an accurate control response.

2.7.5 Sliding Mode Control

As mentioned earlier, the SM control is a type of nonlinear control introduced initially as a means for controlling variable structure systems. The first instances of its application on DC–DC converters are reported in 1983 [6] and 1985 [104]. Since then, there has been a wide interest in the research community concerning its development. The main advantage of the SM control over

---

[3]US Patent 5 278 490, issued in January 1994.

other types of nonlinear control methods is its ease of implementation. This makes it well suited for common DC–DC power regulation purposes. However, the main problem associated with the application of SM control is its variable frequency nature, which causes excessive power losses and electromagnetic interference. Nonetheless, if this problem is properly handled, SM control does have a huge potential in industrial applications.

3

Sliding Mode Control in Power Converters

CONTENTS

3.1   Introduction

The SM control is naturally well suited for the control of variable structure systems. Characterized by switching, power converters are inherently variable structure systems. It is, therefore, appropriate to apply SM control on power converters [105]. Moreover, SM control offers excellent large-signal handling capability, which is important for DC–DC converters. Since the design of conventional pulse-width modulation (PWM) controllers is small-signal based, the converters being controlled operate optimally only for a specific condition [57] and often fail to perform satisfactorily under large parameter or load variations (i.e., large-signal operating condition) [25, 36, 58]. By replacing linear PWM controllers with SM (nonlinear) controllers, power converters can achieve better regulation and dynamical performance for a wider operating range.

In this chapter, we will give an overview of the different aspects of the application of SM control to DC–DC converters. Our discussion starts with a brief review on the state-of-the-art research and development in the area of SM control of DC–DC converters. We will discuss in detail the various problems associated with applying SM control to power converters, focusing particularly on the issues related to the practical implementation of SM controllers.

3.2 Review of Literature

The literature review is organized under several categories of discussions. Our purpose is to provide a historical overview of the major stages of development in this area.

3.2.1 Earliest Works

Application of SM controllers in DC–DC converters was first reported in 1983 by Bilalović *et al.* [6], who demonstrated the feasibility of using SM control in the buck converter. Later, Venkataramanan *et al.* [104] presented a more complete description the application of SM control to all basic second-order DC–DC converters, and introduced the idea of applying the *equivalent control method* of SM control theory to the standard *duty-ratio control* scheme to achieve constant frequency SM controllers.

3.2.2 Higher-Order Converters

After Venkataramanan *et al.* [104] published their work on second-order DC–DC converters, Huang *et al.* [34] experimented the SM controller on a fourth-order Ćuk converter in 1989. This spurred a series of related works on the Ćuk converter. In 1992, Fossas *et al.* [26] examined the audio-susceptibility and load disturbance property of the SM-controlled Ćuk converter. In 1995, Malesani *et al.* [46] systematized the design approach for the SM-controlled Ćuk converter. In 1996, Oppenheimer *et al.* [66] implemented an SM controller on a third-order Ćuk converter. Also in 1996, Mahdavi *et al.* [43] developed the first PWM-based SM-controlled Ćuk converter. Then, in 1997, Cavente *et al.* [8] introduced a method of designing a locally-stable SM controller for the bi-directional coupled-inductor Ćuk converter, and in 1998, provided a thorough analytical evaluation of the converter [49].

At the same time, the work by Huang *et al.* [34] also generated new interests in applying SM control to other types of higher order DC–DC converters. In 1993, Mattavelli *et al.* [52] proposed a general-purpose SM controller which is applicable for both Ćuk and Sepic converters. In 1997, the same group extended the research by proposing a method of deriving the small-signal models of these converters, which allow the investigation of stability and the selection of the control parameters [53]. In 1994, Dominguez *et al.* [19] performed a stability analysis of an SM-controlled buck converter with an input filter. In 2000, Castilla *et al.* [11] presented the design methodology of SM control schemes for quantum resonant converters.

3.2.3   Parallel-Connected Converters

The interest in applying SM control to more complex types of DC–DC converters has also covered the class of parallel-connected DC–DC converters. In 1996, Donoso *et al.* [20] and Shtessel *et al.* [76] proposed the use of SM control to achieve current equalization and output voltage regulation of modular DC–DC converters. In 1998, Shtessel *et al.* [77] suggested that dynamic sliding surface can be employed for better stabilization and control of such converters.

Also in 1998, López *et al.* [41] proposed the use of SM control for interleaving parallel-connected DC–DC converters. However, the illustration was only limited to the buck-type modular converters. Then in 2000, Giral *et al.* [29] demonstrated the application of SM controllers for interleaved boost converters.

In 2002, Mazumder *et al.* [55] combined the concepts of integral-variable-structure and multiple-sliding-surface control to nullify the bus-voltage error and the error between the load currents of the parallel-connected converter modules. In 2004, López *et al.* [42] presented a detailed account on the analysis and design of an SM-controlled parallel-connected boost converter system. The work by López *et al.* [42] is more relevant to practical power electronics, compared to other prior works on basic theory of SM control [20, 55, 76, 77].

3.2.4   Theoretical Works

Due to simplicity and ease of analysis, simple second-order DC–DC converters have been the subjects of investigation in much of the previous work. The main focus was the theoretical derivation of SM control methodologies. These earlier studies have played an important role in the development of practical SM controllers for DC–DC converters.

In 1994, Sira *et al.* [80] proposed the use of an extended linearization method in SM controller design that results in SM controllers that have excellent self-scheduling properties. In 1995, they proposed to incorporate passivity based controllers into an SM-controlled DC–DC converter to enhance its robustness properties [81, 82]. In 1996, Fossas *et al.* [27] presented a new approach for the design of SM-controlled DC–DC converters for generating an AC signal. In 1997, Carrasco *et al.* [10] proposed to incorporate artificial neural networks with SM controllers for power factor correction (PFC) applications.

In 2001, Bock *et al.* [7] proposed a design procedure for the selection of high-pass filter parameters of the SM-controlled bi-directional DC–DC converters. In 2002, Fossas *et al.* [28] applied a second-order SM controller to a buck converter to reduce the chattering. Also in 2002, Shtessel *et al.* [74, 75] proposed two SM control strategies for boost and buck-boost converters: one using the method of stable system center and the other using a dynamic sliding manifold.

In 2003, Vazquez *et al.* [103] proposed a new sliding surface that eliminates the use of a current sensor, and Gupta *et al.* [30] proposed a hybrid SM controller that employs a combined form of voltage and current sliding surfaces to improve robustness. Also in 2003, Sira *et al.* [83] proposed to use hysteresis modulation in an SM controller to achieve a generalized-proportional-integral continuous control of a buck converter. They also presented a tutorial that revisits and evaluates the performances of direct and indirect SM controller schemes, and proposed the use of generalized-proportional-integral control technique to improve the robustness of the system with respect to un-modeled load resistance variations [84].

Apart from developing SM control theory for power converter applications, evaluations of performance and comparisons with other control methods were also reported. In 1997, Raviraj *et al.* [70] made a comparative study of the buck converter's performance when controlled by proportional-integral (PI), SM, and fuzzy logic controllers. They concluded that there are some similarities in the system's behavior between fuzzy logic and SM controllers. In 2002, Cortes *et al.* [17] revisited the work on the SM control of boost converters by comparing and analyzing the performances using different proposed schemes and sliding surfaces. In 2002, Morel *et al.* [59, 60, 61] studied the nonlinear behavior exhibited by the conventional current-mode controlled boost converter and proposed the alternative use of SM controller, to eliminate chaotic behavior of the converter.

3.2.5   Practical Works

A few experimental evaluations of SM-controlled DC–DC converters have been reported in the literature. Most of the reported experimental works, however, focused their attention on performance evaluation rather than on developing design procedures.

In 1999, Escobar *et al.* [23] performed experiments to compare five different control algorithms, including the SM control scheme, on a DC–DC boost power converter. They concluded that nonlinear controllers provide a promising alternative to linear average controller, which performs poorly in tracking time-varying references. Also in 1999, Chiacchiarini *et al.* [16] conducted an experiment to compare the performances of a buck converter controlled by digital and analog SM controllers. They concluded that the SM-controlled system gives consistent responses despite variation in load conditions.

In 2001, Alarcon *et al.* [5] reported the first analog integrated-circuit (IC) SM control prototype for DC–DC converters. They concluded that the megahertz operating range of the controller fits the requirements supported by modern power electronics technologies. In 2003, Ahmed *et al.* [2, 3] first implemented and then provided an experimental evaluation of the dynamic performance of an SM-controlled buck converter. In another paper [4], they also implemented the SM controller on a buck-boost converter using the control desk dSPACE.

3.2.6 Constant Frequency SM Controllers

Some researchers have noted the importance of maintaining a constant switching frequency operation in their SM-controlled converter systems.

In 1992, Cardoso *et al.* [9] proposed several methods of reducing the switching frequency of the SM-controlled DC–DC converters. In 1995, Nguyen *et al.* [62] proposed an adaptive hysteresis type of SM controllers to ensure constant switching frequency. In 1996, they proposed an indirect method of implementing SM controllers in buck converters so that constant switching frequency can be achieved [63]. In 1997, Mahdavi *et al.* [44] proposed a method of deriving PWM-based SM-controlled DC–DC converters that have a constant switching frequency. Later, in 2000, they extended the work by incorporating artificial neural networks into their PWM-based SM controllers [45]. In 2004, Perry *et al.* [69] proposed a digital fuzzy logic SM like controller that has a fixed switching frequency and provides zero steady-state error, and Iannelli *et al.* [35] proposed a method of dithering to maintain finite and constant switching frequency. Also in 2004, Mazumder *et al.* [56] gave an experimental validation of their proposed control scheme [55] for the parallel-connected converters, which not only optimizes the transient and steady-state responses, but also achieves a constant switching frequency at steady state.

SM control has, in most cases, been studied in continuous time. A few attempts to study its discrete-time counterpart have been reported. In 2000, Matas *et al.* [51] proposed a discrete-time SM-controlled boost converter for output voltage tracking, and Orosco *et al.* [67] provided a complete analysis of discrete-time SM-controlled DC–DC converters. They argued that the discrete-time implementation of the SM controller can overcome the inherent drawbacks of variable switching frequency operation in the conventional continuous-time implementation.

3.2.7 Remarks

In concluding the literature survey, the major developments of SM control in DC–DC converters are summarized as follows, along with some comments on the likely directions for current and future research developments.

Firstly, the idea of applying SM (nonlinear) control in high-order converter systems is acceptable by large since linear controllers are incapable of providing good control over such systems. Higher design and implementation costs can be easily justified in such circumstances. On the other hand, the idea of applying SM controllers to basic second-order DC–DC converter topologies is often challenged and is generally unpopular, even within the research community. The main complaint is the conceptual/implementational complexity of the control scheme as compared to existing linear controllers, which are already offering acceptable control properties in such converters for most applications. Moreover, the notion of using relatively higher-cost digital means to implement the SM controllers, as illustrated in many previous attempts, has been deemed unrealistic for commercial applications. Hence, it is important to find out what the benefits and drawbacks of using SM controllers are, as compared to existing linear PWM controllers, assuming that it is possible to implement the former in a comparable form and cost to the latter.

Secondly, it can be concluded that a large part of the previous effort has been devoted to developing the theoretical/mathematical framework of SM control for DC–DC converters. Practical implementation using simple circuitries, however, has been very much neglected due to the conventional belief that the derived SM control schemes can be easily realized in digital forms. Thus, the interest in its development normally halts at the theoretical/mathematical stages. However, considering that if SM controllers are to be implemented for commercial applications, thorough studies of practical problems are necessary. Specifically, the different means of developing analog SM controllers, which operate at a constant switching frequency complying with the usual industrial standard should be explored. Yet, under such design criteria, a contradiction exists between the ideal operation of SM controllers at infinitely high frequency and the practical constraint of physical converter systems that are only allowed to operate at limited frequency ranges. Thus, while aiming at competitive pricing and compatible standard with existing PWM controllers, there must also be considerations on how the non-ideality of finite switching frequency can be compromised in practical SM controllers without sacrificing the large-signal properties of SM control. Essentially, more investigations into such aspects are still required.

In conclusion, before SM controllers can be commercially viable in common DC–DC converter applications, the aforementioned issues should first be overcome. This may prove to be difficult considering that many still regard the work on the application of SM controllers in simple second-order converters as redundant and impractical. Hence, before it is possible to proceed with further investigation, the viability of using SM controllers for simple converters has to be clearly demonstrated. For instance, it has been shown in prior works [5, 16, 52, 90, 91, 92, 93, 94] that SM controllers can be implemented in terms of simple analog circuits.

3.3 Characteristics of SM Control as Applied to DC–DC Converters

"Practically all design methods for variable structure systems are based on deliberate introduction of sliding modes which have played, and are still playing, an exceptional role both in theoretical developments and in practical applications." – extracted from Utkin *et al.* [102].

If we are to elaborate such a reasoning in the context of power electronics, a logical deduction follows that since all power electronics converters are intrinsically variable structured, their control methodologies, be it linear or nonlinear, are all based on some form of SM control. This is true since the objectives of all controllers in the DC–DC converters share the same formula of switching between multiple-structures to achieve a desired output. Ultimately, they are all tracking some kind of sliding surfaces, each uniquely defined by its control strategy, to achieve an equilibrium. Therefore, the difference between conventional control methodologies and the actual SM control methodology can be distinguished by the way in which the controllers are being designed. For conventional control methodologies, the sliding surfaces are indirectly formulated, through the design of the control parameters and controller's type, which are determined by some stability analyses. The controller designers are not informed, nor have they any direct control on how the sliding surfaces are constructed. For the SM control, however, the design procedure starts with the formulation of the sliding surfaces. The controller designers always determine the type of sliding surfaces they desire and this allows them to exercise a direct control over the dynamic response of the system. The way in which the control parameters are chosen is then purely defined by the hitting condition, the existence condition, and the stability condition of the SM control law.

Such insights can be taken into consideration when determining how different control objectives can be achieved in power electronics converters. Following this, we will discuss the various aspects of the SM controllers that are relevant to DC–DC converters.

3.3.1 General Principle of SM Control Implementation

As discussed in Chapter 1, the basic principle of SM control is to design a certain sliding manifold in its control law that will direct the trajectory of the state variables toward a desired operating point. In the case of a single switch DC–DC converter, it is appropriate to have a control law that adopts a switching function such as

$$u = \frac{1}{2}\left(1 + \text{sign}(S)\right) \tag{3.1}$$

where u is the logic state of the converter's power switch and S is the instantaneous state trajectory, which in the case of a second-order controller, is

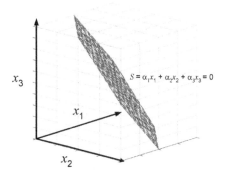

FIGURE 3.1
Graphical representation of the sliding plane in three-dimensional space.

described as

$$S = \alpha_1 x_1 + \alpha_2 x_2 + \alpha_3 x_3 = \boldsymbol{J}^{\mathrm{T}} \boldsymbol{x}, \tag{3.2}$$

with $\boldsymbol{J}^{\mathrm{T}} = [\alpha_1 \ \ \alpha_2 \ \ \alpha_3]$ and α_1, α_2, and α_3 representing the control parameters, also referred to as sliding coefficients. x_1, x_2, and x_3 denotes the desired state feedback variables to be controlled. By enforcing $S = 0$, a sliding plane (manifold) can be obtained. The graphical representation of the sliding plane in a three-dimensional space is illustrated in Fig. 3.1.

From our discussion in Chapter 1, it can be understood that the entire SM control process for the converter will be divided into two phases, namely, the reaching phase and the sliding phase. In the reaching phase, regardless of the starting position, the controller will perform a control decision that will drive the trajectory of the state variables to converge to the sliding plane (see Fig. 3.2(a)) through the compliance of the hitting condition.

When the trajectory is within a small vicinity of the sliding plane, the converter enters the sliding phase where the trajectory is maintained within a small vicinity of the sliding plane, and is concurrently directed toward the desired reference at origin O (see Fig. 3.2(b)) through the compliances of the existence condition and the stability condition.

When the system enters into SM operation, its equivalent trajectory is ideally $S = 0$, which defines the dynamic characteristic of the system, and can be designed by the proper choice of sliding coefficients [1].

3.3.2 Constant Dynamics in Power Converters

Recall that the SM control has the virtue of providing constant system's dynamics during the sliding phase. In this example, the transient dynamics for the converter during the sliding phase can be obtained by equating (3.2) as

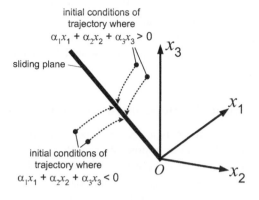

(a) Phase 1

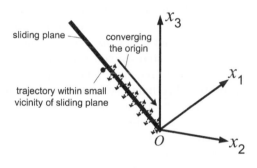

(b) Phase 2

FIGURE 3.2
Graphical representations of state trajectory's behavior in SM control process:
(a) Reaching phase—illustrating trajectory converging the sliding plane irrespective of its initial condition; and (b) Sliding phase—illustrating trajectory
being maintained within a small vicinity of the sliding plane, and concurrently
being directed to converge to the origin O.

$S = 0$, i.e.,

$$\frac{d^2 x_1}{dt^2} + \frac{\alpha_1}{\alpha_2} \cdot \frac{dx_1}{dt} + \frac{\alpha_3}{\alpha_2} \cdot x_1 = 0. \tag{3.3}$$

This property applies only to the sliding phase and not the reaching phase.
Since the time taken to complete the sliding phase is typically much longer
than the reaching phase, it is sufficient to consider only the dynamics of the
sliding phase for the control design.

This is not possible with the conventional type of linear PWM control. The linear-compensation structure of the PWM control allows the dynamics of the system to be characterized only at one specific operation condition, which will result in a changing dynamic condition for different operating conditions. A more thorough comparison of these properties will be given in the later part of this book.

3.3.3   Quasi-Sliding Mode Control in Power Converters

Theoretically, to achieve a near-perfect SM control operation, the system should be operated at the highest possible switching frequency so that the controlled variables can follow closely the reference track to achieve the desired dynamic response and steady-state operation [102]. This requirement for operation at a very high switching frequency, however, challenges the feasibility of applying SM control to DC–DC converters. This is because extreme high-speed switching in DC–DC converters results in excessive switching losses, inductor and transformer core losses, and "electromagnetic interference" noise issues. Hence, for SM control to be applicable to DC–DC converters, their switching frequencies must be constricted within a practical range. Up to now, several methods (hysteresis; constant sampling frequency; constant on-time; constant switching frequency; and limited maximum switching frequency) have been proposed to limit the switching frequency [9].

Nevertheless, the constriction of the SM controller's switching frequency transforms the controller into a quasi-sliding mode (QSM) controller, which operates as an approximation of the ideal SM controller. The consequence of this transformation is the reduction of the system's robustness and the deterioration of the regulation properties. Clearly, the proximity of QSM to the ideal SM controller will be closer as switching frequency tends to infinity. Since all SM controllers in practical DC–DC converters are frequency limited, they are, strictly speaking, QSM controllers. The term SM controller has been customarily adopted to represent QSM controller in many research papers. This will also be adopted in this book.

3.3.4   Conventional Hysteresis-Modulation-Based Implementation

As mentioned in Chapter 1, the conventional method of implementing the SM control is by using a relay of the signum function given in (3.1) along with the trajectory S computed using (3.2). However, as discussed, this direct implementation of the SM control law results in converters that operate at a very high and uncontrolled frequency [102], which makes them unsuitable for many applications. To limit the operating frequency within a controllable range, the hysteresis function–based SM control as described in Chapter 1 is used. This is referred to as the hysteresis-modulation (HM)-based SM control.

The implementation is easily accomplished by refining (3.1) as

$$u = \begin{cases} 1 = \text{`ON'} & \text{when } S > \kappa \\ 0 = \text{`OFF'} & \text{when } S < -\kappa \\ \text{previous state} & \text{otherwise} \end{cases} \tag{3.4}$$

where κ is an arbitrarily small value that will affect the switching frequency of the power converters. The method of designing the magnitude of κ to achieve the desired switching frequency for a particular operating condition will be discussed in the following chapter.

3.4 Fixed-Frequency SM Controller in Power Converters

The imposition of the hysteresis band into the signum function of the switching relay solves only the problem of high-frequency switching of the converters. It does not solve the problem of variable switching frequency operation in the power converters. Power converters with HM-based SM control generally suffer from having a significant variation in their switching frequencies when the input voltage and the output load are varied [52, 90]. This makes the process of designing the converters and their input and output filters difficult. Obviously, designing the filters under a worst-case (lowest) frequency condition will result in oversized filters. Also, the variation of the switching frequency is known to deteriorate the regulation properties of the converters [93]. Moreover, it is well known that switching converters are severe noise generators, and the task of containing noise is much easier with a fixed switching-frequency. This makes it necessary to have the SM controllers designed to operate at a fixed frequency regardless of the operating conditions.

There are basically three approaches in keeping the switching frequency of the HM-based SM controller constant. One approach is to incorporate a constant ramp or timing function directly into the controller [9, 35, 52]. The main advantage of this approach is that the switching frequency is constant under all operating conditions, and can be easily controlled through varying the ramp/timing signal. However, this comes at the expense of additional hardware circuitries, as well as deteriorated transient response in the system's performance. This is because with the superposition of the ramp function upon the SM switching function, what is achieved is a deteriorated form of the original SM control. For these reasons, it is not a good approach for implementing fixed frequency in SM control as it changes the original control characteristic, and we will omit this approach in our subsequent discussion.

The second approach is to include some form of adaptive control into the HM-based SM controller to counteract the switching frequency variation [62, 93]. For line variation, the frequency variation is reduced through

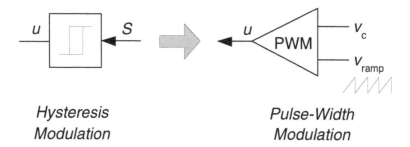

**Hysteresis   Pulse-Width
Modulation   Modulation**

FIGURE 3.3
Simplified hysteresis modulation (HM) and pulse-width-modulation (PWM)
structure.

the adaptive feedforward control, which varies the hysteresis band with the
change of the line input voltage. For load variation, the frequency variation
is reduced through the adaptive feedback control, which varies the control
parameter (i.e., sliding coefficient) with the change of the output load [93].
Conceptually, these methods of adaptive control are more direct, keep the
original characteristic of SM control, and do not suffer from a deteriorated
transient response. A more detailed discussion on how such adaptive SM controllers can be implemented will be given in Chapter 5.

On the other hand, constant switching frequency SM controllers can also
be obtained by employing pulse-width modulation (PWM) instead of HM
[63, 93]. In practice, this is similar to classical PWM control schemes in which
the control signal is compared to the ramp waveform to generate a discrete
gate pulse signal [58]. The advantages are that additional hardware circuitries
are not needed since the switching function is performed by the pulse-width
modulator, and that its transient response is not deteriorated. However, the
implementation is non-trivial in order to preserve the original SM control law,
especially when both the current and voltage state variables are involved.
Hence, this approach is not always implementable for some SM controller
types.

As this is an unconventional and relatively new topic, a short discussion
is provided in the following section.

3.4.1 Pulse-Width Modulation-Based Sliding Mode Con-troller

Figure 3.3 illustrates the idea of the PWM-based SM controller, where PWM
is used in lieu of HM, without destroying the SM control properties. This
requires the relationship of the two control techniques to be established. Two
key results are useful here. First, in SM control, the discrete control input
(gate signal) u can be theoretically replaced by a smooth function known as
the *equivalent control* signal u_{eq}, which can be formulated using the *invariance*

conditions by setting the time differentiation of (3.2) as $\dot{S} = 0$ [102]. Second, at a high switching frequency, the equivalent control is effectively a duty-ratio control [79]. Since a duty ratio is basically also a smooth analytic function of the discrete control pulses in PWM, we can obtain a PWM-based SM controller by translating the equivalent control function to the duty ratio d of the pulse-width modulator, i.e., $d = u_{eq}$.

Interestingly, the derived PWM-based SM controllers can also be viewed as a type of nonlinear state-feedback controllers designed from some nonlinear "per-cycle averaged models" of the converters. However, it should be emphasized that a main difference between the two approaches is that while an "average model" is assumed, the PWM-based SM controller approach, which only performs averaging during the controller implementation, retains much of the converter dynamics. This results in a set of design restriction imposed upon the existence condition, which arises from the instantaneous dynamics of the converter, as required by the SM control theory. Such design restrictions are absent from the conventional nonlinear PWM controller design approach.

A detailed discussion on how the HM-based and the PWM-based SM controllers can be developed will be provided in the following chapters. But first, we will provide an overview to the theory of duty-ratio control and how it can be related to the equivalent control theory of SM control.

3.4.2 Duty-Ratio Control

In conventional PWM control, which is also known as the duty-ratio control, the control input u is switched between 1 and 0 once every switching cycle for a fixed small duration Δ. The time instants at which the switching occurs are determined by the sampled values of the state variables at the beginning of each switching cycle. The duty ratio is then the fraction of the switching cycle in which the control holds the value 1. It is normally a smooth function of the state-variable vector $x(t)$, and is denoted by $d(x)$, where $0 < d(x) < 1$. Then, for each switching cycle interval Δ, the control input u can be written as

$$u = \begin{cases} 1 & \text{for } t \leq \tau < t + d(x)\Delta \\ 0 & \text{for } t + d(x)\Delta \leq \tau < t + \Delta \end{cases} . \tag{3.5}$$

It follows that a system $\dot{x} = f(x, u)$ can be expressed as

$$x(t + \tau) = x(t) + \int_{t}^{t+d(x)\Delta} [f(x(\tau))]\, d\tau + \int_{t+d(x)\Delta}^{t+\Delta} 0\, d\tau. \tag{3.6}$$

The ideal average model of the PWM-controlled system response is obtained by allowing the switching frequency to tend to infinity, i.e., Δ to approach zero. Under such consideration, the above equation becomes

$$\frac{\lim_{\Delta \to 0}[x(t + \Delta) - x(t)]}{\Delta} = \frac{\lim_{\Delta \to 0}\left[\int_{t}^{t+d(x)\Delta} f(x(\tau))d\tau\right]}{\Delta}, \tag{3.7}$$

i.e.,

$$\frac{dx}{dt} = \dot{x} = f(x, d),\tag{3.8}$$

which is referred to as the average PWM-controlled system. Therefore, as the switching frequency tends to infinity, the ideal average behavior of the PWM-controlled system is represented by the smooth response of the system constituted by the duty ratio $d(x)$. It should also be noted that the duty ratio $d(x)$ replaces the discrete function u in the same manner as the equivalent control u_{eq} of the SM control scheme to obtain (1.20). Hence, the relationship $d(x) = u_{\text{eq}}(x)$ is established, which forms the theoretical basis for the implementation of the equivalent control using the PWM control.

3.5   Some Design Guidelines

Since SM control can achieve order reduction, it is typically sufficient to have an SM controller of $(n - 1)$th order for the stable control of an nth order converter. However, if fixed-frequency SM controllers are to be employed, the robustness and regulation properties of the converter system under the order-reduced SM controller will be deteriorated. A good method of alleviating these deteriorations is to introduce an additional integral control variable into the fixed-frequency SM controller. This is known as *integral* or *full-order SM control*, since the SM controller is now of the same order as the converter [102]. The purpose of making the SM controller full-order is to improve the robustness as well as the regulation of the system. Hence, in terms of ease of implementation and control performance, a good option is to adopt an SM controller that employs a linear combination of the system states which has the same order as the converter.

Now, assuming that a full-order sliding mode controller is designed for a basic second-order converter, the control function of such a second-order controller is generally expressed as

$$u = \begin{cases} U^+ & \text{when } S > 0 \\ U^- & \text{when } S < 0 \end{cases}\tag{3.9}$$

where S takes the form as described in (3.2). The task of the designer is to determine the state of U^+/U^- and to select proper values of α_1, α_2, and α_3 such that the controller meets the hitting, existence, and stability conditions for all of the system's operating input and loading conditions.

Step 1: To Meet Hitting Condition

The design of the SM controller to meet the hitting condition is rather straightforward in the case of power converters. Taking the output voltage as the control variable, the state variables of the full-order SM controller to be controlled may be expressed in the following form:

$$\begin{bmatrix} x_1 \\ x_2 \\ x_3 \end{bmatrix} = \begin{bmatrix} V_{\text{ref}} - \beta v_{\text{o}} \\ \frac{d(V_{\text{ref}} - \beta v_{\text{o}})}{dt} \\ \int (V_{\text{ref}} - \beta v_{\text{o}}) dt \end{bmatrix} \tag{3.10}$$

where V_{ref} and βv_{o} denote the reference and sensed instantaneous output voltages, respectively; and x_1, x_2, and x_3 are the *voltage error*, the *voltage error dynamics* (or the rate of change of voltage error), and the *integral of voltage error*, respectively. For the design of the hitting condition, it is sufficient to consider only the immediate state variable x_1, which is predominant in the composition of S during the reaching phase. Apparently, if the sensed output voltage is much lower than the reference voltage, i.e., S is positive, the intuitive switching action required for the compensation is to turn on the power switch so that energy is transferred from the input source to the inductor. Conversely, if the sensed output voltage is much higher than the reference voltage, i.e., S is negative, the intuitive switching action is to turn off the power switch so that energy transfer between the source and the inductor is discontinued. This forms the basis for the formulation of the hitting condition. The resulting control function under this configuration is

$$u = \begin{cases} 1 = \text{'ON'} & \text{when } S > 0 \\ 0 = \text{'OFF'} & \text{when } S < 0 \end{cases}. \tag{3.11}$$

Clearly, the method of ensuring the fulfillment of the hitting condition of the SM controller is closely related to the way in which the switching states of the hysteresis controller are designed. Thus, the same approach may be adopted for ensuring the hitting condition of an SM controller which employs other types of sliding manifold.

Step 2: To Meet Existence Condition

With the switching states U^+/U^- determined, the next stage is to ensure that the selected sliding coefficients α_1, α_2, and α_3 comply with the condition for SM existence. This is possible by inspecting the local reachability condition of the state trajectory, i.e.,

$$\lim_{S \to 0} S \cdot \dot{S} < 0. \tag{3.12}$$

In the case of a buck converter, substitution of the converter's description into the above condition gives

$$0 < LC \frac{\alpha_3}{\alpha_2} \left(V_{\text{ref}} - \beta v_{\text{o}} \right) - \beta L \left(\frac{\alpha_1}{\alpha_2} - \frac{1}{r_{\text{L}} C} \right) i_C + \beta v_{\text{o}} < \beta v_{\text{i}} \tag{3.13}$$

where C, L, and r_L denote the capacitance, inductance, and instantaneous load resistance of the converters, respectively; V_{ref}, v_i, and v_o denote the reference, instantaneous input, and instantaneous output voltages, respectively; β denotes the feedback network ratio; and i_C denotes the instantaneous capacitor current.

Here, $C, L, \beta, V_{\text{ref}}$ are known parameters of the converter system and their exact values can be substituted directly into the inequality for inspection. However, for v_i and r_L, which typically represent a range of input and output operating conditions, it is necessary to consider the boundary points of these operating conditions. The compliance of either the maximum or minimum point of these operating conditions is generally sufficient for ensuring the abidance of the existence condition for the entire range of operating condition. As for v_o and i_C, which are instantaneous state variables, the consideration of the time-varying nature of these components undesirably complicates the design. In the case of designing an SM controller with a static sliding surface, a practical approach is to ensure that the existence condition is met for the steady-state operation [53, 84]. With such considerations, the state variables i_C and v_o can be substituted with their expected steady-state values, i.e., $i_{C(\text{SS})}$ and $v_{o(\text{SS})}$, which can be derived from the design specifications. This ensures the compliance of the existence condition at least in the small region around the origin. Taking into consideration the above conditions, the resulting existence condition is

$$0 < LC\frac{\alpha_3}{\alpha_2}\left(V_{\text{ref}} - \beta v_{o(\text{SS})}\right) - \beta L\left(\frac{\alpha_1}{\alpha_2} - \frac{1}{r_{L(\min)}C}\right)i_{C(\text{SS})} \qquad (3.14)$$
$$+\beta v_{o(\text{SS})} < \beta v_{i(\min)}.$$

The selected sliding coefficients must comply with the stated inequality.

Step 3: To Meet Stability Condition

In addition to complying with the existence condition, the selected sliding coefficients must concurrently satisfy the stability condition. Interestingly, this can be inherently accomplished through the design of the sliding coefficients to meet the desired dynamical property [1].

In our example, the equation relating the sliding coefficients to the dynamic response of the converter during SM operation is

$$\alpha_1 x_1 + \alpha_2 \frac{dx_1}{dt} + \alpha_3 \int x_1 dt = 0. \qquad (3.15)$$

This equation can be rearranged into a standard second-order system form in which the design of the sliding coefficients α_1, α_2, and α_3 can result in the system adopting one of these three possible types of responses: under-damped, critically damped, and over-damped, with a desired convergency rate [91]. Hence, designers can easily select the sliding coefficients based on their converters' response time and voltage overshoot specifications. The selected sliding coefficients must also comply with the existence condition described in the previous section. Similar steps can be adopted for designing controllers of a different sliding manifold.

3.6 Practical Issues in Analog Implementation

Before we conclude this chapter, some practical issues concerning the implementation of SM controllers for DC–DC converters should be noted. Study related to practical implementation is still relatively insufficient, especially in the area of analog implementation.

Issues such as the requirement for constant switching frequency operation in the SM controller and the need to redefine the sliding coefficients to meet practical component constraints have been covered in some previous studies [90, 91, 92, 93, 94]. Moreover, some aspects related to the implementation of SM controllers should be more thoroughly investigated in order to gain sufficient insights for designing practical SM controllers for DC–DC converters.

First, the choice of system's state variables, i.e., voltage, current, their derivatives and/or integrals, is important as it affects the control performance as well as the complexity of the implementation. Some practical questions that should be considered are

1. How many state variables are needed for accomplishing the required control function?
   Comment: Having more state variables would mean that more sensing and/or computation of the state variables are required.

2. What is the choice of the state variables to be controlled?
   Comment: The use of voltage state variables is easier because of the simplicity of implementing voltage sensors. However, the use for their derivatives may involve the use of noise sensitive differentiators. Alternatively, indirect means of sensing such state variables is possible, e.g. dv_o/dt can be found by sensing the current of the output capacitor. Yet, this will lead us to the subsequent question as to whether it is convenient to perform such sensing. On the other hand, an integral term of the controlled variable $\int x$ is often required to reduce the steady-state error of the system. Otherwise, the steady-state error may fail to meet the regulation requirement.

3. Is the location accessible or feasible for performing voltage or current sensing?
   Comment: The sensing of capacitor current to produce the derivative of voltage variable is not always the best option. This is especially true for the case where the capacitor is a filter capacitor. Insertion of a current

sensor in the current flow path will increase the impedance, and therefore deteriorate the filtering process.

4. What types of sensors are required?
   Comment: For the sensing of the filtering capacitors' currents, very low-impedance current transformers are normally required. However, for sensing currents that have DC average components, resistors or sophisticated hall-effect sensors may be required. These may affect the overall efficiency or cost of the DC–DC converters.

Moreover, for the PWM-based SM controllers, the indirect implementation of the original SM control law may result in unexpected complication in the signal computation and is therefore not always implementable for some types of SM controllers. Hence, the choice of the state variables is critical for the successful implementation of an analog SM controller.

Finally, similar to conventional controllers, the physical limitation of the analog devices, e.g. bandwidths, propagation time delays, slew rates, and saturation limits, of the SM controller should be properly noted. They are the key factors affecting the proper operation of the SM controller. More work is still required in this area of investigation.

4

Hysteresis-Modulation-Based Sliding Mode Controllers

CONTENTS

4.1 Introduction

Much of the work in the design of hysteresis-modulation (HM)-based SM controller has been developed from the control's perspective rather than from the circuit's perspective [45, 49, 55, 62, 63, 78, 79, 88, 104], although taking

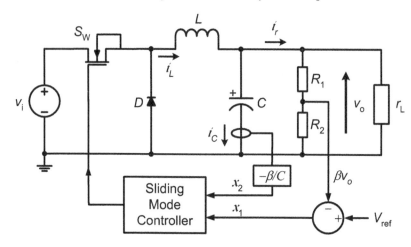

FIGURE 4.1
Basic structure of an SMVC buck converter.

the circuit's perspective would be more useful for practicing circuit design engineers in the physical realization of the controllers. The objective of this chapter is to present the practical design issues pertaining to the implementation of SM controlled converters. For example, the issues of designing the sliding coefficients, hysteresis band, calculation of the switching frequency, and its circuit implementations, etc., will be dealt with in this chapter. A simple and systematic procedure that bridges the gap between the control principle and circuit implementation, which is useful for implementing such controllers is also described. As an illustrative example, the sliding-mode voltage controlled (SMVC) buck converter is employed in the discussion. The approach, however, is generally applicable to all other DC–DC converter types.

4.2   Theoretical Derivation

The complete mathematical derivations of both the ideal and practical converter design of the SMVC buck converter are covered in this section.

4.2.1   Mathematical Model of Buck Converter

To illustrate the underlying principle, the state-space description of the buck converter under SM voltage control, where the control parameters are the output voltage error and the voltage error dynamics (in phase canonical form) [62], is first discussed.

Figure 4.1 shows the schematic diagram of an SMVC buck converter. Here, the *voltage error* x_1 and the *voltage error dynamics* (or the rate of change of voltage error) x_2 under continuous conduction mode, can be expressed as

$$
\begin{aligned}
x_1 &= V_{\text{ref}} - \beta v_{\text{o}} \\
x_2 &= \dot{x}_1 = -\beta \frac{dv_{\text{o}}}{dt} = \frac{\beta}{C}\left(\frac{v_{\text{o}}}{r_{\text{L}}} - \int \frac{uv_{\text{i}} - v_{\text{o}}}{L} dt \right)
\end{aligned}
\tag{4.1}
$$

where C, L, r_{L} are the capacitance, inductance, and instantaneous load resistance, respectively; V_{ref}, v_{i}, and βv_{o} are the reference, instantaneous input, and instantaneous sensed output voltages, respectively; $u = 1$ or 0 is the switching state of power switch S_{W}. Then, by differentiating (4.1) with respect to time, the state-space model can be obtained as

$$
\begin{bmatrix} \dot{x}_1 \\ \dot{x}_2 \end{bmatrix} = \begin{bmatrix} 0 & 1 \\ -\frac{1}{LC} & -\frac{1}{r_{\text{L}}C} \end{bmatrix} \begin{bmatrix} x_1 \\ x_2 \end{bmatrix} + \begin{bmatrix} 0 \\ -\frac{\beta v_{\text{i}}}{LC} \end{bmatrix} u + \begin{bmatrix} 0 \\ \frac{V_{\text{ref}}}{LC} \end{bmatrix}.
\tag{4.2}
$$

The graphical representation of the respective substructure of the system with $u = 1$ and $u = 0$, for different starting (x_1, x_2) conditions, is shown in Fig. 4.2. It can be seen that when $u = 1$, the phase trajectory for any arbitrary starting position on the phase plane will converge to the equilibrium point $(x_1 = V_{\text{ref}} - \beta v_{\text{i}}, x_2 = 0)$ after some finite time period. Similarly, when $u = 0$, all trajectories converge to the equilibrium point $(x_1 = V_{\text{ref}}, x_2 = 0)$. These characteristics will be exploited for the design of the SM voltage controller.

4.2.2   Design of an Ideal SM Voltage Controller

In SM control, the controller employs a switching function to determine its input states to the system [102]. For an SM voltage controller, the switching function u can be determined from the control parameters x_1 and x_2 using the state trajectory computation:

$$
S = \alpha x_1 + x_2 = \boldsymbol{J}\boldsymbol{x}
\tag{4.3}
$$

where α is the control parameter (termed as sliding coefficient) to be designed; $\boldsymbol{J} = [\alpha, 1]$; and $\boldsymbol{x} = [x_1, x_2]^T$. By enforcing $S = 0$, a *sliding line* with gradient α can be obtained. The purpose of this sliding line is to serve as a boundary to split the phase plane into two regions. Each of this region is specified with a switching state to direct the phase trajectory toward the sliding line. When the phase trajectory reaches and tracks the sliding line toward the origin, the system is considered to be stable, i.e., $x_1 = 0$ and $x_2 = 0$.

The specification of the switching state for each sector in the case of a second-order system like the buck converter can be graphically performed by observing the behavior of the trajectories in Fig. 4.3, which is a combination of the two plots in Fig 4.2. It can be observed that if the phase trajectory is at any arbitrary position above the sliding line ($S = 0$), e.g. point M, $u = 1$

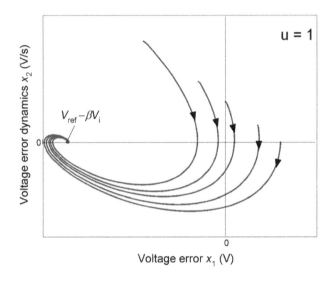

(a) $u = 1$

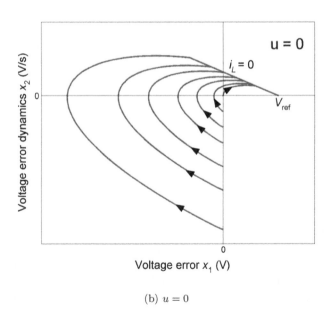

(b) $u = 0$

FIGURE 4.2

Phase trajectories of the substructure corresponding to (a) $u = 1$ and (b) $u = 0$ for different starting (x_1, x_2) positions.

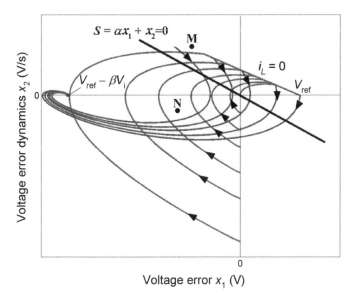

FIGURE 4.3
Combined plot of phase trajectories of the substructure corresponding to both $u = 1$ and $u = 0$ for different (x_1, x_2) starting positions.

must be employed so that the trajectory is directed toward the sliding line. Conversely, when the phase trajectory is at any position below sliding line, e.g. point N, $u = 0$ must be employed for the trajectory to be directed toward the sliding line. This forms the basis for the control law:

$$u = \begin{cases} 1 = \text{`ON'} & \text{when } S > 0 \\ 0 = \text{`OFF'} & \text{when } S < 0 \end{cases}. \tag{4.4}$$

Complying with the *hitting condition* [88], the system trajectory will eventually reach the sliding line, and the control law in (4.4) provides the general requirement that the trajectories will be driven toward the sliding line. However, there is no assurance that the trajectory can be maintained on this line. To ensure that the trajectory is maintained on the sliding line, the existence condition, which is derived from *Lyapunov's second method* [85] to determine asymptotic stability, must be obeyed [102, 104]:

$$\lim_{S \to 0} S \cdot \dot{S} < 0. \tag{4.5}$$

Thus, by substituting the time derivative of (4.3), the condition for SM control to exist is

$$\dot{S} = \begin{cases} J\dot{x} < 0 & \text{for } 0 < S < \xi \\ J\dot{x} > 0 & \text{for } -\xi < S < 0 \end{cases} \tag{4.6}$$

where ξ is an arbitrarily small positive quantity. Substituting (4.2) and (4.4) into (4.6), the inequalities become

$$\lambda_1 = \left(\alpha - \frac{1}{r_L C}\right) x_2 - \frac{1}{LC} x_1 + \frac{V_{\text{ref}} - \beta v_i}{LC} < 0$$

$$\lambda_2 = \left(\alpha - \frac{1}{r_L C}\right) x_2 - \frac{1}{LC} x_1 + \frac{V_{\text{ref}}}{LC} > 0 \qquad (4.7)$$

where

$$\lambda_1 = \boldsymbol{J}\dot{\boldsymbol{x}} \quad \text{for} \quad 0 < S < \xi$$
$$\lambda_2 = \boldsymbol{J}\dot{\boldsymbol{x}} \quad \text{for} \quad -\xi < S < 0. \qquad (4.8)$$

The above conditions are depicted in Fig. 4.4 for the two respective situations: (a) $\alpha > \frac{1}{r_L C}$ and (b) $\alpha < \frac{1}{r_L C}$. In both figures, Region 1 represents $\lambda_1 < 0$ and Region 2 represents $\lambda_2 > 0$. SM operation will only be valid on the portion of the sliding line, $S = 0$, that covers both Regions 1 and 2. In this case, this portion is within A and B, where A is the intersection of $S = 0$ and $\lambda_1 = 0$; and B is the intersection of $S = 0$ and $\lambda_2 = 0$. Since the trajectory will slide to the origin only when it touches $S = 0$ within AB, it will go beyond the sliding line if the trajectory lands outside AB (as shown in Fig. 4.4(a)). This results in an overshoot in the voltage response when $\alpha > \frac{1}{r_L C}$. Hence, for the practical condition that $0 \leq v_o \leq v_i$, the system trajectory will be bounded within the region $V_{\text{ref}} - \beta v_i < x_1 < V_{\text{ref}}$. Taking this into account, the maximum existence region will occur when $\alpha = \frac{1}{r_L C}$.

Furthermore, by manipulating (4.3), we have

$$\alpha x_1 + \dot{x}_1 = 0 \quad \Rightarrow \quad x_1(t) = x_1(t_0)e^{-\alpha(t-t_0)} \qquad (4.9)$$

where t_0 is any point in time and $x_1(t_0)$ is the voltage error at t_0. From this, it can be understood that the choice of α in the SM design is more than controlling the existence region. It also controls the dynamic response of the system with a first-order time constant of

$$\tau = \frac{1}{\alpha}. \qquad (4.10)$$

Hence, to ensure that α is high enough for fast dynamic response and low enough to maintain a large existence region, it is sufficient to set

$$\alpha = \frac{1}{r_L C}. \qquad (4.11)$$

On the other hand, if a faster dynamic response is desired, α can be reduced, as long as the existence condition of the system is met. For this purpose, it is necessary to take into consideration the operating condition of the converter. Now, assuming that the converter is to operate in an input

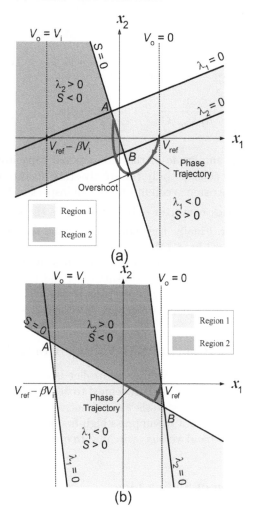

FIGURE 4.4

Regions of existence of SM in phase plane: (a) $\alpha > \frac{1}{r_L C}$ and (b) $\alpha < \frac{1}{r_L C}$.

voltage range $V_{i(min)} \leq v_i \leq V_{i(max)}$ and an output load resistance range $R_{L(min)} \leq r_L \leq R_{L(max)}$. Then, equation (4.18) can be rewritten as

$$\frac{V_{od} - V_{i(min)}}{L} < -\left(\alpha - \frac{1}{R_{L(max)}C}\right)\left|\hat{i_C}\right| \tag{4.12}$$

$$\frac{V_{od}}{L} > \left(\alpha - \frac{1}{R_{L(max)}C}\right)\left|\hat{i_C}\right| \tag{4.13}$$

or more explicitly as

$$
\alpha < \begin{cases} \dfrac{V_{i(min)} - V_{od}}{L\left|\hat{i}_C\right|} + \dfrac{1}{R_{L(max)}C} & \text{when } V_{i(min)} < 2V_{od} \\[4mm] \dfrac{V_{od}}{L\left|\hat{i}_C\right|} + \dfrac{1}{R_{L(max)}C} & \text{otherwise} \end{cases} , \qquad (4.14)
$$

where $\hat{i}_C$ is the peak amplitude of the bidirectional capacitor current flow. Since it is sufficient to consider only the steady-state operating condition for the compliance of the existence condition, we can inspect (4.14) by substituting $\left|\hat{i}_C\right| = \left|\tilde{i}_L\right|$ (for buck converter only), where $\tilde{i}_L$ is the steady-state ripple current of the inductor. Finally, the choice of α should comply with the existence condition described in (4.14).

It should, however, be noted that α must be a positive quantity to achieve system stability. This can be analyzed using (4.8) and (4.11), which by substitution of a negative α quantity, will result in, respectively, a trajectory that moves away from the phase plane origin and an x_1 that does not tend to zero.

4.2.3   Design of a Practical SM Voltage Controller

In this section, a practical SM voltage controller is considered. The sliding line defined in the previous section is redefined to accommodate for hardware limitations. Additionally, a hysteresis band is introduced to the sliding line as a form of frequency control to suppress high-frequency switching. The relationship of hysteresis band versus switching frequency of the SMVC buck converter is derived.

4.2.3.1   Redefinition of Sliding Line

As previously mentioned, an SM controller requires the continuous assessment of the parameters x_1 and x_2. By substituting (4.1) and (4.11) into (4.3), we have

$$
S = K_{p1}(V_{ref} - \beta v_o) + K_{p2}i_C \qquad (4.15)
$$

where $K_{p1} = \frac{1}{r_L C}$ and $K_{p2} = -\frac{\beta}{C}$. In the above equation, the terms $(V_{ref} - \beta v_o)$ and i_C are the feedback state variables from the converter that should be amplified by gains K_{p1} and K_{p2}, respectively, before a summation is performed. This, from a practical perspective, does generate a problem. Noting that capacitance C in DC–DC converters is usually in the microfarad (μF) range, its inverse term will be significantly higher than β and r_L. Hence, the overall gains K_{p1} and K_{p2} will become too high for practical implementation. If forcibly implemented, the feedback signals may be driven into saturation, thereby causing (4.15) to provide unreliable information for the control.

In view of that, it is simpler to reconfigure the switching function to the following description:

$$S = \frac{C}{\beta}\alpha x_1 + \frac{C}{\beta}x_2 = \boldsymbol{Qx} \tag{4.16}$$

where $\boldsymbol{Q} = [\frac{C}{\beta}\alpha \ \ \frac{C}{\beta}]$; and $\boldsymbol{x} = [x_1 \ \ x_2]^T$. From (4.1) and (4.11), we get

$$S = \frac{1}{\beta r_L}(V_{ref} - \beta v_o) - i_C. \tag{4.17}$$

Thus, the practical implementation of S becomes independent of C, thereby reducing the amplification of the feedback signals. With this sliding line, the conditions for SM control to exist are

$$\lambda_1 = \left(\frac{C}{\beta}\alpha - \frac{1}{\beta r_L}\right)x_2 - \frac{1}{\beta L}x_1 + \frac{V_{ref} - \beta v_i}{\beta L} < 0$$

$$\lambda_2 = \left(\frac{C}{\beta}\alpha - \frac{1}{\beta r_L}\right)x_2 - \frac{1}{\beta L}x_1 + \frac{V_{ref}}{\beta L} > 0 \tag{4.18}$$

where

$$\lambda_1 = \boldsymbol{Q\dot{x}} \quad \text{for} \quad 0 < S < \xi$$

$$\lambda_2 = \boldsymbol{Q\dot{x}} \quad \text{for} \quad -\xi < S < 0. \tag{4.19}$$

An interesting point here is that although there is modification to the λ equations, the maximum existence region will still occur at $\alpha = \frac{1}{r_L C}$. In addition, the response time is still maintained at $\tau = \frac{1}{\alpha}$.

4.2.3.2 Introduction of Hysteresis Band

Ideally, a converter will switch at infinite frequency with its phase trajectory moving along the sliding line when it enters SM operation (see Fig. 4.5(a)). However, in the presence of switching imperfections, such as switching time constant and time delay, this is not possible. The discontinuity in the feedback control will produce a particular dynamic behavior in the vicinity of the sliding surface known as chattering (see Fig. 4.5(b)) [21, 68, 101, 102].

If the chattering is left uncontrolled, the converter system will become self-oscillating at a very high switching frequency corresponding to the chattering dynamics. As mentioned in the earlier chapter, this is undesirable as high switching frequency will result in excessive switching losses, inductor and transformer core losses, and "electromagnetic interference" noise issues.[1] Furthermore, since chattering is introduced by the imperfection of controller ICs, gate driver and power switches, it is difficult to predict the exact switching frequency. Hence, the design of the converter and the selection of the components will be difficult.

---

[1]Most PWM power supplies are designed to operate with switching frequencies between 40 kHz and 200 kHz.

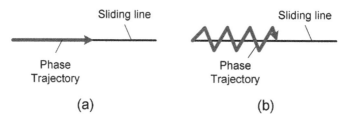

FIGURE 4.5
Phase trajectory for (a) ideal SM operation; (b) actual SM operation with chattering.

To solve these problems, the control law in (4.4) is redefined as

$$u = \begin{cases} 1 = \text{`ON'} & \text{when } S > \kappa \\ 0 = \text{`OFF'} & \text{when } S < -\kappa \\ \text{unchanged} & \text{otherwise} \end{cases} \quad (4.20)$$

where κ is an arbitrarily small value. The reason for introducing a hysteresis band with the boundary conditions $S = \kappa$ and $S = -\kappa$ is to provide a form of control to the switching frequency of the converter. This is a method commonly employed to alleviate the chattering effect of SM control [85]. With this modification, the operation is altered such that if the parameters of the state variables are such that $S > \kappa$, switch S_W of the buck converter will turn on. Conversely, it will turn off when $S < -\kappa$. In the region $-\kappa \le S \le \kappa$, S_W remains in its previous state. Thus, by introducing a region $-\kappa \le S \le \kappa$ where no switching occurs, the maximum switching frequency of the SM controller can be controlled. This alleviates the effect of chattering. In addition, it is now possible to control the frequency of the operation by varying the magnitude of κ.

4.2.3.3 Calculation of Switching Frequency

To control the switching frequency of the converter, the relationship between the hysteresis band, κ, and the switching frequency, f_S, must be known.

Figure 4.6 shows the magnified view of the phase trajectory when it is operating in SM. $\mathbf{f}^-$ and $\mathbf{f}^+$ are the vectors of state variable velocity for $u = 0$ and $u = 1$, respectively. It was previously derived that [101]

$$\Delta t_1 = \frac{2\kappa}{\nabla S \cdot \mathbf{f}^-}$$
$$\Delta t_2 = \frac{-2\kappa}{\nabla S \cdot \mathbf{f}^+} \quad (4.21)$$

where Δt_1 is the time taken for vector $\mathbf{f}^-$ to move from position x to y; and Δt_2 is the time taken for vector $\mathbf{f}^+$ to move from position y to z. By

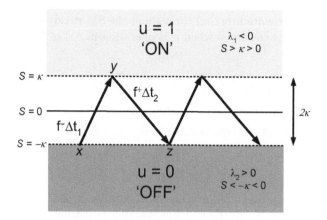

FIGURE 4.6
Magnified view of the phase trajectory in SM operation.

substituting in (4.21)

$$\nabla S \cdot f = \sum_{i=1}^{n} \frac{\partial S}{\partial x_i} \frac{dx_i}{dt} = \frac{dS}{dt} = \dot{S} \qquad (4.22)$$

where

$$f = \begin{cases} f^- & \text{for } u = 0 \\ f^+ & \text{for } u = 1 \end{cases},$$

we have

$$\Delta t_1 = \frac{2\kappa}{\dot{S}_{u=0}} = \frac{2\kappa}{\lambda_2}$$

$$\Delta t_2 = \frac{-2\kappa}{\dot{S}_{u=1}} = \frac{-2\kappa}{\lambda_1}. \qquad (4.23)$$

Further substitution of (4.18) into (4.23) results in

$$\Delta t_1 = \frac{2\kappa}{\left(\frac{C}{\beta}\alpha - \frac{1}{\beta r_L}\right) x_2 - \frac{1}{\beta L} x_1 + \frac{V_{\text{ref}}}{\beta L}}$$

$$\Delta t_2 = \frac{-2\kappa}{\left(\frac{C}{\beta}\alpha - \frac{1}{\beta r_L}\right) x_2 - \frac{1}{\beta L} x_1 + \frac{V_{\text{ref}} - \beta v_i}{\beta L}}. \qquad (4.24)$$

Therefore, the time period during steady state, i.e., $x_1 = 0$, for one cycle in which the phase trajectory moves from position x to z is equivalent to $T = \Delta t_1 + \Delta t_2$, which gives

$$T = \frac{-2\kappa v_i}{\left[\left(\frac{C}{\beta}\alpha - \frac{1}{\beta r_L}\right) x_2\right]^2 L + \left[\left(\frac{C}{\beta}\alpha - \frac{1}{\beta r_L}\right) x_2\right] (2v_o - v_i) + \frac{v_o(v_o - v_i)}{L}}. \qquad (4.25)$$

Since the cycle is repeated (cyclic) throughout the SM steady-state operation, the frequency of the converter when it is operating in SM is $f_S = \frac{1}{T}$, and can be expressed as

$$f_S = \frac{\left[\left(\frac{C}{\beta}\alpha - \frac{1}{\beta r_L}\right)x_2\right]^2 L + \left[\left(\frac{C}{\beta}\alpha - \frac{1}{\beta r_L}\right)x_2\right](2v_o - v_i) + \frac{v_o(v_o - v_i)}{L}}{-2\kappa v_i}. \quad (4.26)$$

Using $\alpha = \frac{1}{r_L C}$, the above equation becomes

$$f_S = \frac{v_o\left(1 - \frac{v_o}{v_i}\right)}{2\kappa L}. \quad (4.27)$$

Considering that v_i and v_o are non-constant parameter consisting of, respectively, DC signals of $\overline{V_i}$ and $\overline{V_o}$ and time-varying perturbations of $\tilde{v_i}$ and $\tilde{v_o}$, we can resolve (4.27) into

$$f_S = \overline{f_S} + \tilde{f_S} \quad (4.28)$$

where $\overline{f_S}$ represents the steady-state (nominal cyclic) switching frequency and $\tilde{f_S}$ represents the ac varying (perturbed) frequency of the converter. This can be done by first separating the variables v_i, v_o, and f_S from (4.27) into their steady-state and small-signal terms:

$$
\begin{aligned}
v_i &= \overline{V_i} + \tilde{v_i}; \\
v_o &= \overline{V_o} + \tilde{v_o}; \\
f_S &= \overline{f_S} + \tilde{f_S}.
\end{aligned} \quad (4.29)
$$

Substituting (4.29) into (4.27), we have

$$
\begin{aligned}
f_S &= \frac{(\overline{V_o} + \tilde{v_o})\left(1 - \frac{\overline{V_o} + \tilde{v_o}}{\overline{V_i} + \tilde{v_i}}\right)}{2\kappa L} \\
&= \frac{\overline{V_o} + \tilde{v_o} - \frac{\overline{V_o}^2 + 2\overline{V_o}\tilde{v_o} + \tilde{v_o}^2}{\overline{V_i} + \tilde{v_i}}}{2\kappa L}.
\end{aligned} \quad (4.30)
$$

Since the output voltage ripple is very small, i.e., $\tilde{v_o}$ is very small, the cross term of $\tilde{v_o}$ can be neglected. Hence, (4.30) becomes

$$f_S = \frac{\overline{V_o} + \tilde{v_o} - \frac{\overline{V_o}^2 + 2\overline{V_o}\tilde{v_o}}{\overline{V_i} + \tilde{v_i}}}{2\kappa L}. \quad (4.31)$$

From (4.31), the steady-state equation and the small-signal equation can be obtained as

$$\overline{f_S} = \frac{\overline{V_o}\left(1 - \frac{\overline{V_o}}{\overline{V_i}}\right)}{2\kappa L} \quad (4.32)$$

$$\tilde{f_S} = \frac{\tilde{v_o}\left(1 - \frac{2\overline{V_o}}{\overline{V_i} + \tilde{v_i}}\right)}{2\kappa L}. \quad (4.33)$$

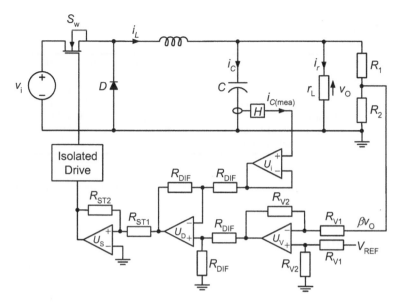

FIGURE 4.7
A standard SMVC buck converter.

This indicates that if there are significantly small variations in the input and output voltages, i.e., $\overline{V_i} \gg \widetilde{v}_i$ and $\overline{V_o} \gg \widetilde{v}_o$, the converter will be operating at a steady-state switching frequency of $\overline{f_S}$ with very little ac frequency perturbation, i.e., $\overline{f_S} \gg \widetilde{f}_S$. Since only the nominal steady-state operating conditions are considered in the controller's design, only (4.32) is required for the design of the steady-state switching frequency of the converter.

4.3 A Standard Design Procedure

A standard SMVC buck converter module is proposed in this section, along with a *step-by-step* design procedure for practical implementation. A design example is provided for illustration.

4.3.1 Standard SMVC Converter Model

Figure 4.7 shows the proposed SMVC buck converter. The SM controller comprises basically a differential amplifier circuit U_V, a voltage follower circuit U_i, a difference amplifier circuit U_D, and a non-inverting Schmitt Trigger circuit U_S. As in most conventional schemes, the feedback sensing network for v_o is provided by the voltage divider circuit, R_1 and R_2. In addition, a low resis-

TABLE 4.1

Specifications of buck converter.

Description	Parameter	Nominal Value
Input voltage	v_i	24 V
Minimum capacitance	C_{min}	4 μF
Critical inductance	L_{crit}	34.28 μH
Desired switching frequency	f_{Sd}	200 kHz
Nominal load resistance	$R_{L(nom)}$	6 Ω
Desired output voltage	V_{od}	12 V

tance current transformer is placed in series with the filter capacitor to obtain the capacitor current, i_C.

4.3.2 Design Steps

Our discussion starts with the assumption that the converter's parameters are known and are given in Table 4.1.

These parameters are calculated on the basis that the converter is to be operated in continuous conduction mode for $v_i = 13$ V to 30 V and $i_r = 0.5$ A to 4 A. The maximum peak-to-peak ripple voltage is 50 mV.

4.3.2.1 Step 1: Current Sensing Gain—H

The current sensing gain, H, is set at a value such that the measured capacitor current, $i_{C(mea)}$ is equal to the actual capacitor current, i_C.

4.3.2.2 Step 2: Voltage Divider Network—β

Setting reference voltage $V_{ref} = 3.3$ V, β is calculated using the expression:

$$\beta = \frac{V_{ref}}{V_{od}} = \frac{3.3}{12} = 0.275 \text{ V/V}. \qquad (4.34)$$

Also, R_1 and R_2 are related by

$$R_2 = \frac{\beta}{1-\beta} R_1. \qquad (4.35)$$

Choosing R_1 as 870 Ω, we get $R_2 = 330 \ \Omega$.

4.3.2.3 Step 3: Gain of Differential Amplifier—U_V

From (4.17), the gain required for the amplification of the signal $(V_{ref} - \beta v_o)$ is $\frac{1}{\beta r_L}$. Setting $r_L = R_{L(nom)}$, R_{V1} and R_{V2} are then related by

$$R_{V1} = (\beta R_{L(nom)}) R_{V2}. \qquad (4.36)$$

Choosing $R_{V2} = 20$ kΩ, we get $R_{V1} = 33$ kΩ. Also, the resistors, R_{DIF}, in the difference amplifier circuit U_D, are all 10 kΩ.

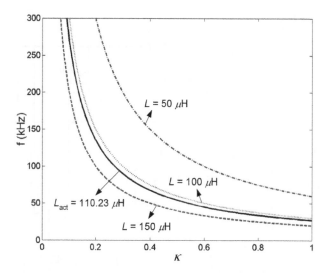

FIGURE 4.8
Calculated κ values for inductances of 50 μH, 100 μH, 110.23 μH (actual inductance), and 150 μH at switching frequencies of up to 300 kHz.

4.3.2.4 Step 4: Calculation of Hysteresis Band—κ

The parameter of the hysteresis band, κ, can be obtained from the re-arranged form of (4.32), i.e.,

$$\kappa \;=\; \frac{V_{\mathrm{od}}\left(1 - \frac{V_{\mathrm{od}}}{v_{\mathrm{i}}}\right)}{2f_{\mathrm{Sd}}L} \tag{4.37}$$

where v_{i}, V_{od}, and f_{Sd} are the nominal parameters of the converter. A plot giving the calculated κ values for different inductances and switching frequencies is shown in Fig. 4.8.

The actual inductance, L_{act}, used in the design is 110.23 μH. It should be noted that $L_{\mathrm{act}} \geq L_{\mathrm{crit}}$ for continuous conduction mode. Thus, substituting $L = 110.23$ μH into (4.37), κ is calculated as 0.136 VHz^{-1}H^{-1}.[2]

---

[2]The presence of hysteresis band in the switching function introduces an error in the output voltage. This is commonly due to the finite propagation time delay and limited slew rates of practical hardware components, as well as the inability to exercise a completely non-DC-biased hysteresis band. It is important to limit the hysteresis band κ to a small value to minimize this error. On the other hand, if κ is too small, it may be very sensitive to the change of κ (i.e., high df/dκ). A good value is to set κ in the range 0.1 VHz^{-1}H^{-1} $\leq \kappa \leq$ 0.2 VHz^{-1}H^{-1}. Otherwise, a different inductance may be used to keep κ within the range.

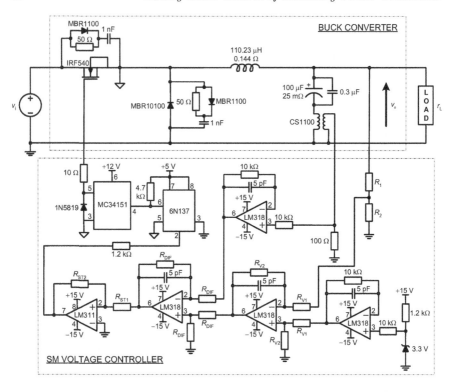

FIGURE 4.9
Full schematic diagram of the SMVC buck converter prototype.

4.3.2.5　Step 5: Design of Schmitt Trigger—U_S

The description of the non-inverting Schmitt Trigger U_S is expressed as

$$2\kappa \;=\; \frac{R_{ST1}}{R_{ST2}}(V_{CC}{}^+ - V_{CC}{}^-) \tag{4.38}$$

where $V_{CC}{}^+$ and $V_{CC}{}^-$ are, respectively, the positive and negative voltage supplies to Schmitt Trigger U_S. Then, the setting of κ for the hysteresis band can be performed by adjusting the ratio of R_{ST1} and R_{ST2}, using the rearranged form of (4.38), i.e.,

$$R_{ST2} = \frac{R_{ST1}(V_{CC}{}^+ - V_{CC}{}^-)}{2\kappa}. \tag{4.39}$$

In our example, choosing R_{ST1} as 110 Ω, resistor R_{ST2} is set as 12 kΩ.

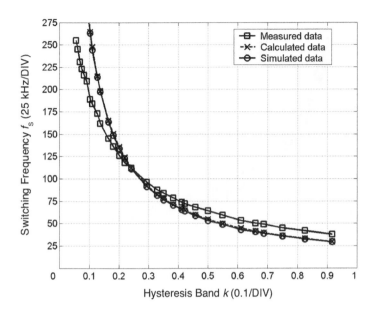

FIGURE 4.10
Calculated, simulated, and experimentally measured average switching frequencies $\overline{f_S}$ at nominal operating condition $v_i = 24$ V and $r_L = 6\ \Omega$ for different hysteresis band κ settings.

4.4 Experimental Results

The performance of the SMVC buck converter that has been designed using the procedure described in the previous section is evaluated in this section. The full schematic diagram of the experimental prototype is shown in Fig. 4.9.

4.4.1 Verification of Design Equation

Figure 4.10 shows the converter's average switching frequency $\overline{f_S}$ for different κ values at nominal operating condition $v_i = 24$ V and $r_L = 6\ \Omega$ that are obtained from calculation, simulation, and measurement. Specifically, the calculation is performed using the proposed design equation (4.32) and the simulation is carried out in Matlab/Simulink using the circuit expression of the proposed controller. Basically, the simulation and experimental data are in good agreement with the calculated data. The small discrepancy between the experimental data and those obtained from calculation and simulation is

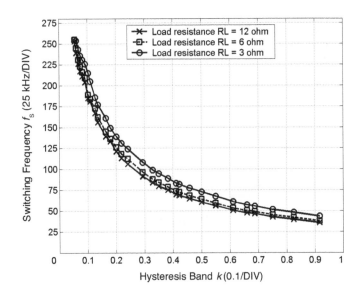

FIGURE 4.11
Experimentally measured average switching frequency $\overline{f_S}$ at $v_i = 24$ V and $r_L = 3\ \Omega$, $6\ \Omega$, and $12\ \Omega$ for different hysteresis band κ settings.

mainly due to component tolerances and the finite time delay of the practical circuits. In practice, fine tuning of κ is required to achieve the desired $\overline{f_S}$.

Figure 4.11 shows the measured average switching frequency $\overline{f_S}$ and Fig. 4.12 shows the measured average output voltage $\overline{V_o}$ against κ for load resistances $r_L = 3$, 6, and 12 Ω. From the figures, $\overline{f_S}$ is notably higher with a lower r_L, and $\overline{V_o}$ is lower with a lower r_L. Additionally, for small κ, i.e., at a high switching frequency, the voltage regulation is tighter and more accurate. By setting $0.1\ \mathrm{VHz}^{-1}\mathrm{H}^{-1} \leq \kappa \leq 0.2\ \mathrm{VHz}^{-1}\mathrm{H}^{-1}$ in the design, we limit the voltage accuracy within an error of ± 0.12 V (i.e., $< 1\%$ of V_{od}).

4.4.2   Steady-State Performance

In our experiment, a controller that is accurately set to operate at a switching frequency of 200 kHz has been used. Specifically, by replacing R_{ST2} with a 16 kΩ resistor, we set $\kappa = 0.1\ \mathrm{VHz}^{-1}\mathrm{H}^{-1}$.

Figure 4.13 shows an example of the output voltage ripple, inductor current, and switching state waveforms of the SMVC buck converter at steady-state operation. Performing to the design expectation, the converter operates at an average switching frequency of $\overline{f_S} = 199$ kHz, with small frequency fluctuations, and the output voltage ripple $\tilde{v}_o$ (without considering the ring-

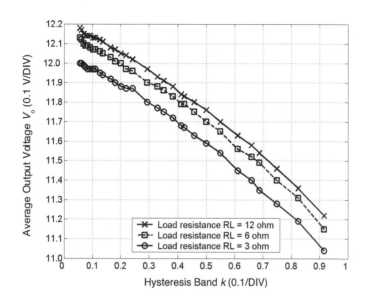

FIGURE 4.12

Experimentally measured average output voltage $\overline{V_o}$ at $v_i = 24$ V and $r_L = 3\ \Omega$, $6\ \Omega$, and $12\ \Omega$ for different hysteresis band κ settings.

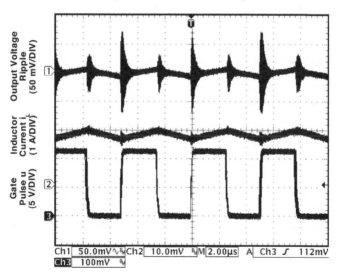

FIGURE 4.13

Waveforms of $\tilde{v}_o$, i_L, and u at steady-state operation under nominal operating conditions $v_i = 24$ V and $r_L = 6\ \Omega$.

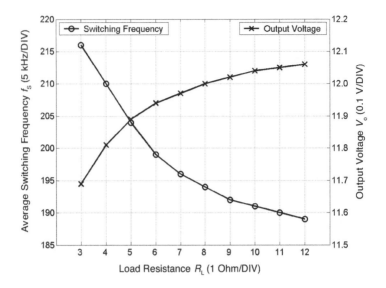

FIGURE 4.14
Experimentally measured values of average switching frequency $\overline{f_S}$ and average output voltage $\overline{V_o}$ for different load resistance r_L.

ing oscillation) is around 10 mV (i.e., $< 0.1\%$ of V_{od}), under the nominal operating condition $v_i = 24$ V and $r_L = 6$ Ω.

4.4.3   Load Variation

Figure 4.14 gives the measured $\overline{f_S}$ and $\overline{V_o}$ for load resistance 3 $\Omega \leq r_L \leq$ 12 Ω. It can be concluded that voltage regulation of the converter is robust to load variation, with only a 0.37 V deviation in $\overline{V_o}$ for the entire load range, i.e., load regulation $\frac{d\overline{V_o}}{dr_L}$ averages at 0.04 V/Ω. The experimental results also indicate that there is a change of switching frequency when the load varies. From the figure, the variation of the switching frequency with respect to load resistance $\frac{df_S}{dr_L}$ averages at -3.0 kHz/Ω. Theoretically, the expression for $\frac{df_S}{dr_L}$ can be obtained by differentiating (4.26) with respect to r_L, i.e.,

$$\frac{df_S}{dr_L} = \frac{Lx_2{}^2}{\beta^2 \kappa v_i} \left(\frac{1}{r_L{}^3} - \frac{C\alpha}{r_L{}^2} \right) + \frac{0.5 - \frac{v_o}{v_i}}{\beta \kappa r_L{}^2} x_2. \tag{4.40}$$

4.4.4   Line Variation

The experimentally measured $\overline{f_S}$ and $\overline{V_o}$ for input voltage range 13 V $\leq v_i \leq$ 30 V are shown in Fig. 4.15. It can be observed that both $\overline{V_o}$ and $\overline{f_S}$ in-

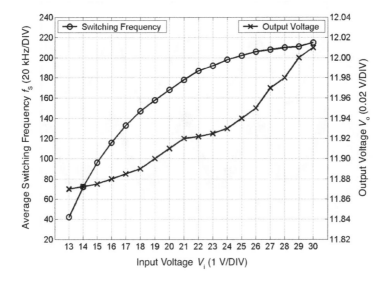

FIGURE 4.15
Experimentally measured values of average switching frequency $\overline{f_S}$ and average output voltage $\overline{V_o}$ for different input voltage v_i.

crease with increasing v_i. The line regulation $\frac{d\overline{V_o}}{dv_i}$ averages at 0.82%. The variation of the switching frequency with respect to input voltage $\frac{d\overline{f_S}}{dv_i}$ averages at 10.176 kHz/V. Theoretically, the expression for $\frac{d\overline{f_S}}{dv_i}$ can be obtained by differentiating (4.27) with respect to v_i, i.e.,

$$\frac{df_S}{dv_i} = \frac{v_o{}^2}{2\kappa L v_i{}^2}. \tag{4.41}$$

Figure 4.16 shows the output voltage ripple waveform of the converter operating at nominal load when v_i is sinusoidally varied from 17.5 V to 29.0 V at a frequency of 100 Hz. This was performed to test the robustness of the converter against a slowly varying input voltage. It is found that the maximum peak-to-peak output voltage is around 235 mV, i.e., the input voltage ripple rejection is −33.8 dB at 100 Hz. The converter has displayed adequate control performance against audio-susceptibility.

4.4.5 α Variation

The dynamic behavior of the converter corresponding to different sliding co-efficients α is also investigated. Figures 4.17(a)–4.17(f) show the output waveforms of the operation with load resistance that alternates between $r_L = 12\ \Omega$ and $r_L = 3\ \Omega$ for the controller with sliding coefficients: (a) $\alpha = \frac{0.5}{R_{L(nom)}C}$; (b)

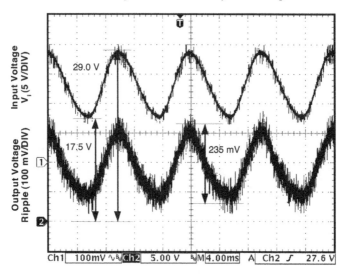

FIGURE 4.16
Waveforms of v_i and $\tilde{v}_o$ under the operating condition whereby v_i is sinusoidally varied from 17.5 V to 29 V at a frequency of 100 Hz, and $r_L = 6\ \Omega$.

TABLE 4.2
Dynamic behavior of the experimental SMVC converter for various α.

Sliding Coefficient Setting	Maximum Voltage Ripple	Settling Time
$\alpha = \frac{0.5}{R_{L(nom)}C}$ $\Rightarrow R_{V1} = 33\ k\Omega;\ R_{V2} = 10\ k\Omega$	±150 mV (±1.25 % of V_{od})	80 ms
$\alpha = \frac{1}{R_{L(nom)}C}$ $\Rightarrow R_{V1} = 33\ k\Omega;\ R_{V2} = 20\ k\Omega$	±100 mV (±0.83 % of V_{od})	60 ms
$\alpha = \frac{2}{R_{L(nom)}C}$ $\Rightarrow R_{V1} = 33\ k\Omega;\ R_{V2} = 40\ k\Omega$	±50 mV (±0.41 % of V_{od})	40 ms
$\alpha = \frac{10}{R_{L(nom)}C}$ $\Rightarrow R_{V1} = 33\ k\Omega;\ R_{V2} = 200\ k\Omega$	±20 mV (±0.17 % of V_{od})	20 ms
$\alpha = \frac{100}{R_{L(nom)}C}$ $\Rightarrow R_{V1} = 33\ k\Omega;\ R_{V2} = 2\ M\Omega$	±1 mV (±0.08 % of V_{od})	1 ms
$\alpha = \frac{500}{R_{L(nom)}C}$ $\Rightarrow R_{V1} = 33\ k\Omega;\ R_{V2} = 10\ M\Omega$	±1 V (±8.33 % of V_{od})	Instable

$\alpha = \frac{1}{R_{L(nom)}C}$; (C) $\alpha = \frac{2}{R_{L(nom)}C}$; (d) $\alpha = \frac{10}{R_{L(nom)}C}$; (e) $\alpha = \frac{100}{R_{L(nom)}C}$; and (f) $\alpha = \frac{500}{R_{L(nom)}C}$, respectively. The results are summarized in Table 4.2.

It can be noted that as α increases, the dynamic response of the converter improves with shorter settling time and smaller overshoots. This is in good agreement with our theory that dynamic response improves with increasing α. However, when α is too high, the converter is unstable (see Fig. 4.17(f)).

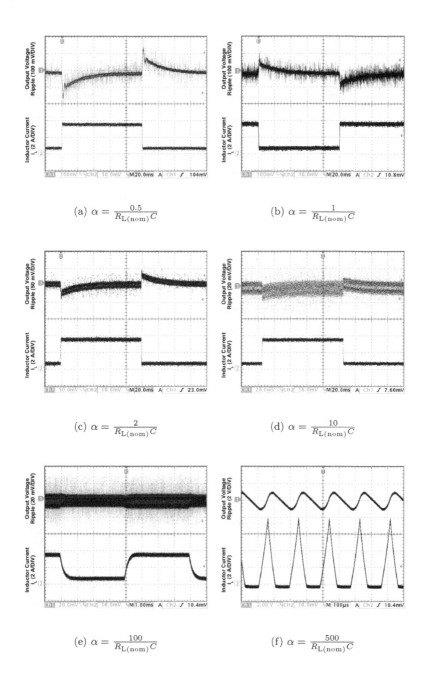

FIGURE 4.17
Output voltage ripple $\widetilde{v}_o$ and inductor current i_L waveforms under step load change alternating between $r_L = 12\ \Omega$ and $r_L = 3\ \Omega$ for different α values.

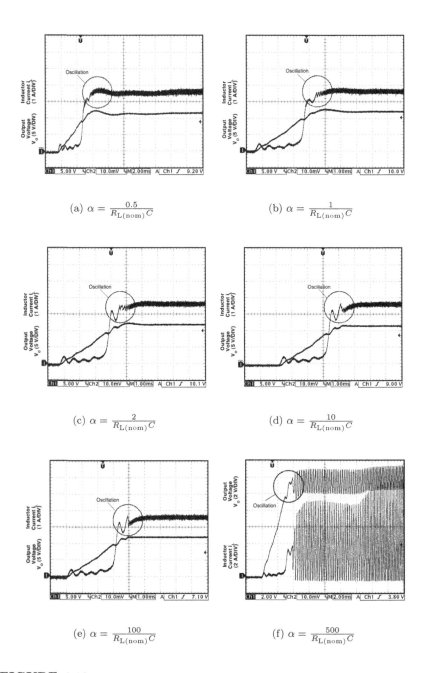

(a) $\alpha = \dfrac{0.5}{R_{\mathrm{L(nom)}}C}$

(b) $\alpha = \dfrac{1}{R_{\mathrm{L(nom)}}C}$

(c) $\alpha = \dfrac{2}{R_{\mathrm{L(nom)}}C}$

(d) $\alpha = \dfrac{10}{R_{\mathrm{L(nom)}}C}$

(e) $\alpha = \dfrac{100}{R_{\mathrm{L(nom)}}C}$

(f) $\alpha = \dfrac{500}{R_{\mathrm{L(nom)}}C}$

FIGURE 4.18
Output voltage v_{o} and inductor current i_{L} waveforms under nominal operating
conditions $v_{\mathrm{i}} = 24$ V and $r_{\mathrm{L}} = 3$ Ω during startup for α values.

This is due to the distortion of the control signal caused by the saturation of the amplified feedback signal (refer to the discussion related to (4.15)).

Figures 4.18(a)–4.18(f) show the output waveforms of the startup operation for the controller with the same set of sliding coefficients. The purpose is to examine the oscillatory behavior of the response due to α, which exercises direct influence on the existence region in the control. Consistent with theory, the amplitude of oscillation increases with the increasing α. Since the magnitude of the oscillations will be high for heavy loads, the effect of oscillation due to α should be taken into consideration when designing high-power converters.

4.4.6 ESR Variation

The converter is also subjected to an equivalent series resistance (ESR) variation test. The aim is to investigate the effect of the output capacitor's ESR on the switching behavior of the converter. Figures 4.19(a)–4.19(c) illustrate the experimental waveforms of the converter with the same output capacitance, for different ESR values (25 mΩ, 125 mΩ, and 225 mΩ), operating under nominal load conditions $v_i = 24$ V and $r_L = 6$ Ω. As expected, output voltage ripples are higher with larger values of ESR. However, the variation of ESR does not influence the capacitor current. In all cases, the peak-to-peak capacitor current remains at 0.36 A, while the average switching frequency is at around 200 kHz. Hence, it is evident that the output capacitor's ESR has little influence on the switching frequency.

4.5 Further Discussion

This section discusses the merits and drawbacks of the proposed standard SMVC converter as compared to conventional current-mode and voltage-mode controlled converters. Suggestions to alleviate the various drawbacks are also provided.

4.5.1 Advantages

The main advantage of the SMVC converter is the simplicity in the controller's design and implementation. Unlike conventional current-mode and voltage-mode controllers, which require special techniques (e.g., pole placement method) to *estimate* their controllers' gain parameters, the SM voltage controller's parameters can be *precisely* calculated from simple mathematical equations.

Furthermore, since the SMVC controller is designed from the large-signal converter model, it is stable and robust to large parameter, line, and load

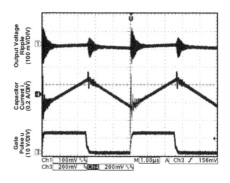

(a) ESR = 25 mΩ

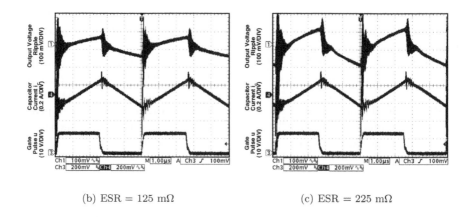

(b) ESR = 125 mΩ   (c) ESR = 225 mΩ

FIGURE 4.19
Experimental waveforms of output voltage ripple $\tilde{v}_o$, capacitor current i_C, and the generated gate pulse u for SMVC converter with 100 μF filter capacitor of (a) ESR = 25 mΩ; (b) ESR = 125 mΩ; and (c) ESR = 225 mΩ, at a constant load resistance r_L = 6 Ω.

variations. This is also a major advantage over conventional current-mode and voltage-mode controllers, which often fail to perform satisfactorily under parameter or large load variations because they are designed from the linearized small-signal converter models [70].

4.5.2 Disadvantages

One major disadvantage of the SMVC converter is that it has a non-zero steady-state voltage error. This is due to the adoption of the phase canonical form in the design, which makes the controller a proportional-derivative (PD) type of feedback controller [70], and the presence of hysteresis band in the switching function, which being non-zero in its average value also introduces an error in the output voltage [88].

Another disadvantage of the SMVC converter is that its steady-state switching frequency is affected by the line and load variations (refer to Figs. 4.14 and 4.15). For line variation, this can be understood from (4.32) where with preset design parameters L, κ, and $\overline{V_o}$, a deviation in the input supply, $\overline{V_i}$, will result in a change in the steady-state DC switching frequency, $\overline{f_S}$. Specifically, $\overline{f_S}$ increases as $\overline{V_i}$ increases. For load variation, the change in switching frequency has two main causes. First, the imperfect feedback loop causes a small steady-state error in the output voltage which in turn causes small deviation of the switching frequency from its nominal value (see (4.32)). Second, there is always mismatch between the nominal load and the operating load. This can be understood from (4.26) where with preset control parameter $\alpha = \frac{1}{R_{L(nom)}C}$ for a certain nominal load $R_{L(nom)}$, a change in operating load r_L from its nominal value will lead to a frequency deviation not given by (4.27).

4.5.3 Possible Solutions

The steady-state voltage error can be easily eliminated by converting the controller into a proportional-integral-derivative (PID) type through the introduction of an integrator to process the voltage error signal [52, 88]. Such controllers will be discussed in detail in Chapter 11.

One possible method of maintaining a small switching frequency variation against the change in the line input is by introducing an adaptive feedforward hysteresis band control that varies the hysteresis band κ with the change of v_i. In practice, this can be accomplished by imposing a variable power supply, $V_{CC}{}^+$ and $V_{CC}{}^-$, which changes with v_i variation, to power the Schmitt Trigger circuit.

An adaptive feedback controller that varies the parameter, α, with the change of load r_L, can be incorporated to maintain the switching frequency of the converter against load variation. The idea is to adaptively maintain the operating status at $\alpha = \frac{1}{r_L C}$ for all load conditions. As will be explained in the following chapter, such a scheme can be used to improve the system's performances, and also maintain the validity of (4.27) so that f_S becomes independent of r_L. Hence, the effect of frequency variation caused by the mismatch between the nominal and the operating load is eliminated.

Details of the design and derivation of such adaptive feedforward and feedback control schemes will be addressed in the next chapter.

5

Hysteresis-Modulation-Based Sliding Mode Controllers with Adaptive Control

CONTENTS

5.1 Introduction

As discussed in the previous chapter, conventional hysteresis-modulation (HM)-based sliding mode (SM) controlled converters generally suffer from significant switching frequency variation when the input voltage and output load are varied [52]. This is undesirable as it leads to oversized filters in the converters, deteriorates the regulation properties of the converters, and complicates "electromagnetic interference" suppression. Hence, it is very much preferred to operate the converters at a constant switching frequency.

As discussed, there are three possible approaches in keeping the switching frequency of the SM controller constant. One approach is to incorporate a constant ramp or timing function directly into the HM-based SM controller [9, 35, 52]. This approach is fairly well addressed in the literature and is effectively a variant of SM control.

The other approach is to employ pulse-width modulation (PWM) instead of HM [63, 93]. In practice, this is similar to classical PWM control schemes in which the control signal is compared with the ramp waveform to generate a discrete gate pulse signal [58]. This approach will be discussed in the subsequent chapters.

The third approach of keeping the switching frequency of the SM controller constant, which is the topic of discussion in this chapter, is through the use of some form of adaptive control to contain the switching frequency variation of the HM-based SM controller [62]. For line variation, an adaptive feedforward control that varies the hysteresis band in the hysteresis modulator of the SM controller in the event of any change of the line input voltage is employed. For load variation, an adaptive feedback controller that varies the control parameter (i.e., sliding coefficient) according to the change of the output load is employed.

The purpose of this chapter is to present a thorough discussion of the problem of switching frequency variation due to the deviation of operating conditions, and the effectiveness of the adaptive solutions in alleviating the problem. In addition, methods of implementing the adaptive control strategies will be discussed.

5.2   Examination of Conventional HM-Based SM Controlled Converters

The system adopted in this discussion is the same SMVC buck converter used in the previous chapter (see Fig. 4.1). For convenience, the mathematical model previously developed is restated in this section.

5.2.1   Mathematical Model

According to (4.2), the state-space model of the SMVC buck converter is

$$\begin{bmatrix} \dot{x}_1 \\ \dot{x}_2 \end{bmatrix} = \begin{bmatrix} 0 & 1 \\ -\frac{1}{LC} & -\frac{1}{r_L C} \end{bmatrix} \begin{bmatrix} x_1 \\ x_2 \end{bmatrix} + \begin{bmatrix} 0 \\ -\frac{\beta v_i}{LC} \end{bmatrix} u + \begin{bmatrix} 0 \\ \frac{V_{ref}}{LC} \end{bmatrix}. \tag{5.1}$$

From (4.17) and (4.20), the control law of the SMVC buck converter is

$$u = \begin{cases} 1 = \text{'ON'} & \text{when } S > \kappa \\ 0 = \text{'OFF'} & \text{when } S < -\kappa \\ \text{unchanged} & \text{otherwise} \end{cases} \tag{5.2}$$

where

$$S = \frac{1}{\beta r_L}(V_{ref} - \beta v_o) - i_C. \tag{5.3}$$

Also, the width of the hysteresis band κ is a fixed parameter that can be determined using

$$\kappa = \frac{V_{od}\left(1 - \frac{V_{od}}{V_{i(nom)}}\right)}{2 f_{Sd} L} \tag{5.4}$$

with f_{Sd}, V_{od}, and $V_{i(nom)}$ representing the desired steady-state switching frequency, the desired output voltage, and the nominal input voltage, respectively. Note that the above equation is valid only if the sliding coefficient is set as $\alpha = \frac{1}{R_{L(nom)}C}$, and the converter is operating under the nominal load resistance $R_{L(nom)}$. Otherwise, if $\alpha \neq \frac{1}{R_{L(nom)}C}$ or the load resistance differs from $R_{L(nom)}$, the actual switching frequency will differ slightly from the desired frequency f_{Sd}. Furthermore, it has also been shown that under this configuration, the conditions for SM control to exist are

$$
\begin{aligned}
\lambda_1 &= \left(\frac{C}{\beta}\alpha - \frac{1}{\beta r_L}\right) x_2 - \frac{1}{\beta L}x_1 + \frac{V_{ref} - \beta v_i}{\beta L} < 0 \\
\lambda_2 &= \left(\frac{C}{\beta}\alpha - \frac{1}{\beta r_L}\right) x_2 - \frac{1}{\beta L}x_1 + \frac{V_{ref}}{\beta L} > 0.
\end{aligned} \tag{5.5}
$$

5.2.2 Problems Identification

It is generally known that SM controllers using a hysteresis type of modulation suffer from frequency variation when the operating conditions differ from their nominal conditions. This has been experimentally proven, as illustrated in the previous chapter.

5.2.2.1 Experimental Observation

Figure 5.1 shows the experimental data of the buck converter system described in the previous chapter, with specifications as shown in Table 4.1, operating under different input voltages v_i (left) and different load resistances r_L (right). It can be seen that both the output voltage and switching frequency increase with an increasing input voltage. This behavior can be explained in terms of a general form of (5.4), i.e.,

$$f_S = \frac{v_o\left(1 - \frac{v_o}{v_i}\right)}{2\kappa L}, \tag{5.6}$$

which suggests that with preset design parameters, inductor L and hysteresis band κ, and the assumption that the output voltage is held at some constant value v_o, a deviation in the actual input supply v_i will result in a change in the actual switching frequency f_S. This in turn affects the regulation of the output voltage.

As for load variation, it can be seen that the switching frequency decreases while the output voltage increases with an increasing load resistance.

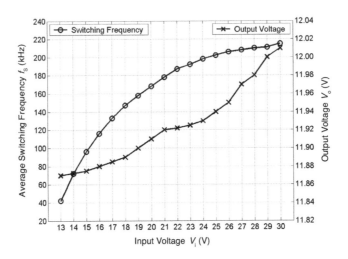

(a) Varying input voltage

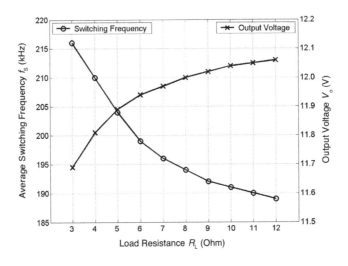

(b) Varying load resistance

FIGURE 5.1
Experimentally measured values of average switching frequency $\overline{f_S}$ and average output voltage $\overline{V_o}$ for (a) different values of input voltage v_i and (b) different values of load resistance r_L.

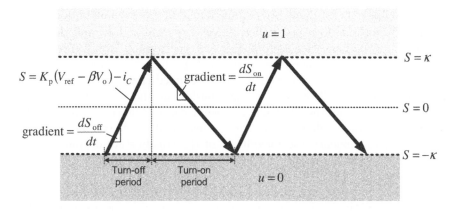

FIGURE 5.2
Switching mechanism of the HM-based SMVC buck converter.

Two causes can be identified for the change in switching frequency. First, the imperfect feedback loop causes a small steady-state error in the output voltage, i.e., $v_o \neq V_{od}$, which in turn causes a small deviation of the switching frequency from its nominal value, i.e., $f_S \neq f_{Sd}$ (see (5.6)). Hence, a change in load will lead to a small change in output voltage and consequently a small change in frequency. Second, in practice, the nominal load almost always differs from the operating load. As previously discussed, equation (5.4) and hence (5.6), will not be strictly true if $\alpha \neq \frac{1}{R_{L(nom)}C}$ or the load resistance differs from $R_{L(nom)}$.

5.2.2.2 Analytical Explanation

The controller operates according to (5.2) and (5.3), which may be graphically represented as shown in Fig. 5.2. Here, the controller generates a turn-on signal $u = 1$ if $S > \kappa$, and a turn-off signal $u = 0$ if $S < -\kappa$. The parameter S is a continuous signal computed using (5.3). By substituting $i_C = i_L - i_r$, where i_L and i_r are, respectively, the inductor and load currents, S can be expressed as

$$S = K_p(V_{ref} - \beta v_o) + \frac{v_o}{r_L} - \int \frac{uv_i - v_o}{L} dt. \tag{5.7}$$

Close examination of this equation and the switching mechanism reveals that in the steady state, the terms $K_p(V_{ref} - \beta v_o)$ and $\frac{v_o}{r_L}$, which contain only DC information, will cancel out the DC component of the term $\int \frac{uv_i - v_o}{L} dt$, leaving its ac component. This means that during steady-state operation, it is effectively the ac component of the term $\int \frac{uv_i - v_o}{L} dt$ that controls the behavior of the trajectory of S. Thus, only a change in v_o or v_i can affect the steady-state switching frequency.

Now, assume that v_o is constant. With $u = 1$ (i.e., the switch is turned on), if v_i is high, the gradient $\frac{dS_{on}}{dt}$ will be high, and the turn-on period of the switch will be short. Conversely, if v_i is low, $\frac{dS_{on}}{dt}$ will be low, and the turn-on period will be long. However, with $u = 0$, the term v_i is nulled from the expression $\int \frac{uv_i - v_o}{L} dt$. Therefore, neither the gradient $\frac{dS_{off}}{dt}$ nor the turn-off period is affected by v_i. Hence, it can be concluded that with increasing v_i, the turn-on period will be shorter, and considering that there is no change in the turn-off period, the switching frequency will be higher. Conversely, the switching frequency will be lower for smaller v_i. Also, since the switching frequency does influence the magnitude of the steady-state output voltage error caused by the imperfect feedback loop, the output voltage is affected by line variation.

However, unlike v_i, which has a direct control over the gradient of the trajectory of S, the load r_L has an indirect and small influence over it. Although r_L appears in (5.7), it is important to recall that in the steady state, the effective ac term in the expression $\int \frac{uv_i - v_o}{L} dt$ does not include r_L. Hence, variation of r_L does not directly affect the trajectory of S. Instead, it is the steady-state error in the output voltage caused by the imperfect feedback loop, which depends on r_L, that leads to the change in the gradient. It should be noted that in this case, both the turn-on and turn-off periods are influenced by r_L. However, since the output voltage is well regulated, the change in its steady-state error will be small. This explains why load changes have only a small effect on the switching frequency. Specifically, as r_L increases, v_o will increase, the gradient of trajectory of S will decrease, and the frequency will decrease (see (5.4)).

5.2.3   Possible Solutions

As mentioned earlier, there are three possible approaches to alleviating the problem of frequency variation caused by line and load variations. The simplest approach is to fix the switching frequency by incorporating a constant ramp or timing function directly into the controller [9, 52, 63], for example, by superposing a constant ramp signal into the switching function [52]; or by including a constant switching frequency circuit in the switching function [9]. This approach is not preferred as it is a deteriorated form of the original SM control. The second approach is by replacing the HM with PWM [63]. However, this approach may not always be implementable for some SM controller types.

Hence, it may be better to consider an alternative approach to solving this problem, that is, to incorporate some form of adaptive control into the HM-based SM controller. The method of restricting the switching frequency variation against line variation can be performed by introducing an adaptive feedforward control that varies the hysteresis band in the hysteresis modulator of the SM controller with the change of the line input voltage. For load variation, this can be performed by incorporating an adaptive feedback controller that varies the control parameter (i.e., sliding coefficient) with the change

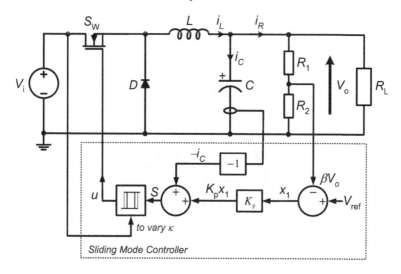

FIGURE 5.3
Basic structure of an adaptive feedforward SMVC buck converter.

of the output load. In the subsequent parts of this chapter, we will describe these adaptive control strategies, their means of implementation, and the main results.

5.3 Adaptive Feedforward Control Scheme

5.3.1 Theory

To keep the switching frequency fixed against line variation, an adaptive feedforward control scheme that varies the hysteresis band in the hysteresis modulator of the SM controller in the event of any change of the line input voltage must be introduced. Figure 5.3 shows the basic structure of the adaptive feedforward SMVC buck converter.

The operation of the adaptive feedforward variable hysteresis band is illustrated in Fig. 5.4, which shows the trajectory of S for one switching cycle of the steady-state operation. Here, the terms $S_{\text{off(min)}}$ and $S_{\text{on(min)}}$ represent the trajectory of S when the input voltage is minimum for, respectively, the turn-off period $T_{\text{off(min)}}$ and turn-on period $T_{\text{on(min)}}$. Similarly, $S_{\text{off(max)}}$ and $S_{\text{on(max)}}$ represent the trajectory of S when the input voltage is maximum, for respectively the turn-off period $T_{\text{off(max)}}$ and turn-on period $T_{\text{on(max)}}$. Also, κ_{min} and κ_{max} represent the required width of the hysteresis band for maintaining the same switching frequency at minimum and maximum input voltage. Recall

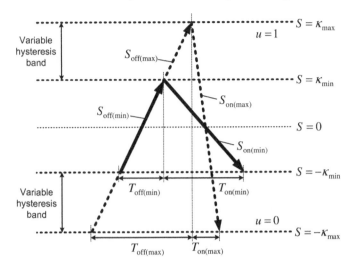

FIGURE 5.4
Operating mechanism of the adaptive feedforward variable hysteresis band.

that when $u = 0$, v_i does not affect the gradient of trajectory S. Hence, the gradient of trajectory $S_{\text{off(min)}}$ and that of trajectory $S_{\text{off(max)}}$ are equivalent. But when $u = 1$, the gradient of trajectory $S_{\text{off(min)}}$ will be smaller than that of trajectory $S_{\text{off(max)}}$. Therefore, it is possible to obtain a hysteresis band of κ_{max} that produces a switching time period $T_{\text{max}} = T_{\text{off(max)}} + T_{\text{on(max)}}$ when the input voltage is maximum, and is equivalent to the switching time period $T_{\text{min}} = T_{\text{off(min)}} + T_{\text{on(min)}}$ that employs the hysteresis band of κ_{min} when the input voltage is minimum. Finally, the input voltage is sensed and the hysteresis band is adjusted accordingly to maintain a certain desired switching frequency.

The real-time computation of the hysteresis band for different input voltages can be performed using a general form of (5.4), where the input voltage term is now v_i, instead of $V_{i\text{(nom)}}$, i.e.,

$$\kappa = \frac{V_{\text{od}}\left(1 - \frac{V_{\text{od}}}{v_i}\right)}{2f_{\text{Sd}}L}. \tag{5.8}$$

Simulation[1] has been performed to verify the concept. Figure 5.5 shows the simulated data for the SMVC buck converter, with and without incorporating the adaptive feedforward control scheme. It can be seen that the adoption of the adaptive feedforward control scheme reduces the variation of $\overline{f_S}$ for the input voltage range $18 \text{ V} \leq v_i \leq 30 \text{ V}$ from $\pm 35\%$ to within $\pm 5\%$ of f_{Sd} (200 kHz).

---

[1]The simulation is performed using Matlab/Simulink. The step size taken for all simulations is 10 ns.

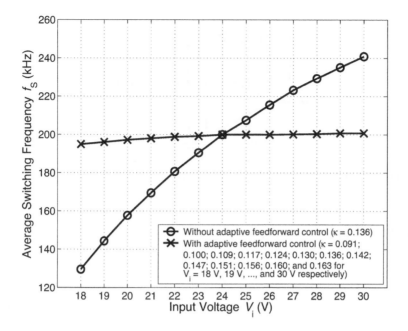

FIGURE 5.5
Simulated data of average switching frequency $\overline{f_S}$ of the SMVC buck converter, with and without the incorporation of the adaptive feedforward control scheme, operating under line variation with input voltage $18\text{ V} \le v_i \le 30\text{ V}$ and at $r_L = 6\ \Omega$.

5.3.2 Implementation Method

Several methods of varying the hysteresis band of the hysteresis modulator are possible. In the case of employing the Schmitt trigger as the hysteresis modulator, the hysteresis band can be adjusted by changing the resistor gain ratio $\frac{R_{ST2}}{R_{ST1}}$, or by adjusting the power supply $V_{CC}{}^+/V_{CC}{}^-$. For this discussion, the latter is chosen.

Figure 5.6 shows the schematic of the non-inverting Schmitt Trigger that can be used for the purpose. The hysteresis band of this circuit is given by

$$2\kappa = \frac{R_{ST1}}{R_{ST2}}(V_{CC}{}^+ - V_{CC}{}^-). \tag{5.9}$$

Equating with (5.8) gives

$$V_{CC}{}^+ - V_{CC}{}^- = \frac{R_{ST2}}{R_{ST1}}\frac{V_{od}}{f_{sd}L}\left[1 - \frac{V_{od}}{V_i}\right]. \tag{5.10}$$

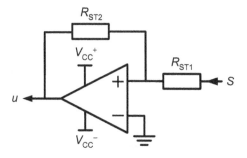

FIGURE 5.6
Schematic of a standard non-inverting Schmitt trigger circuit.

Also, V_{CC}^+ and V_{CC}^- should be

$$V_{\text{CC}}^+ = \frac{1}{2}G_S\left[1 - \frac{V_{\text{od}}}{v_i}\right];$$

$$V_{\text{CC}}^- = -\frac{1}{2}G_S\left[1 - \frac{V_{\text{od}}}{v_i}\right]; \qquad (5.11)$$

where $G_S = \frac{R_{\text{ST2}}}{R_{\text{ST1}}}\frac{V_{\text{od}}}{f_{\text{sd}}L}$. In practice, this variable power supply, $V_{\text{CC}}^+/V_{\text{CC}}^-$, can be obtained through simple analog computation.

5.4 Adaptive Feedback Control Scheme

5.4.1 Theory

The problem of variable switching frequency as load varies is caused by the difference between the operating load resistance and the nominal load resistance used in the controller design, i.e., $r_{\text{L}} \neq R_{\text{L(nom)}}$. When this occurs, equation (5.4), which assumes that the sliding coefficient is chosen as $\alpha = \frac{1}{R_{\text{L(nom)}}C}$ and that the converter is operating with the nominal load resistance $R_{\text{L(nom)}}$, becomes invalid. This results in the deviation of the switching frequency from the nominal value. Thus, a logical solution is to ensure that equation (5.4) is valid for all operating conditions. This can be accomplished by making the sliding coefficient adaptive. Instead of simply fixing it at $\alpha = \frac{1}{R_{\text{L(nom)}}C}$, the sliding coefficient is designed to be load dependent: $\alpha = \frac{1}{r_{\text{L}}C}$, i.e., a change in the operating load resistance r_{L} will immediately change α. Such a system has been proposed in [89] to improve system's performances. Here, it can be used as a means to also maintain the validity of (5.6) so that f_{S} becomes independent of r_{L}. Hence, the effect of frequency variation caused by the mismatch between the nominal and the operating load is alleviated. Note that although

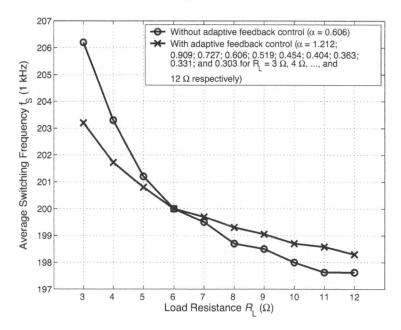

FIGURE 5.7

Simulated data of average switching frequency $\overline{f_S}$ of the SMVC buck converter, with and without the incorporation of the adaptive feedback control scheme, operating under load variation with load resistance $3\ \Omega \le r_L \le 12\ \Omega$ and at $v_i = 24$ V.

the convergence rate of the system is affected by the adaptive feedback control, its stability is preserved since $\alpha > 0$.

Figure 5.7 shows the simulated data for the SMVC buck converter, with and without the incorporation of the adaptive feedback control scheme. It can be seen that the adoption of the adaptive feedback control scheme reduces the variation of $\overline{f_S}$ for the load resistance range $3\ \Omega \le r_L \le 12\ \Omega$ from $\pm 3.1\%$ to within $\pm 1.6\%$ of f_{Sd} (200 kHz).

5.4.2 Implementation Method

By making the sliding coefficient adaptive, i.e., $\alpha = \frac{1}{r_L C}$, SM control equation (5.7) becomes

$$S = \frac{1}{\beta r_L}\left(V_{\text{ref}} - \beta v_o\right) - i_C. \tag{5.12}$$

Clearly, the computation of the control signal S requires the measurement of all involving variables in the equation. However, since it is not possible to

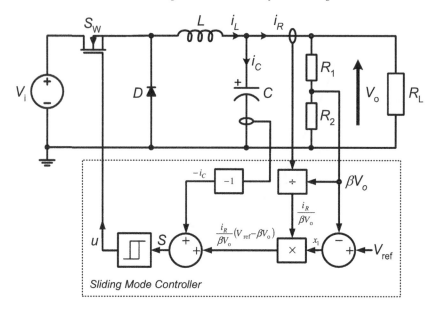

FIGURE 5.8
Basic structure of an adaptive feedback SMVC buck converter.

measure resistance directly, the relationship:

$$r_\mathrm{L} = \frac{v_\mathrm{o}}{i_r} \tag{5.13}$$

is exploited to obtain the instantaneous load resistance. Hence, the adaptive feedback control scheme is practically implemented using the equation

$$S = \frac{i_r}{\beta v_\mathrm{o}}(V_\mathrm{ref} - \beta v_\mathrm{o}) - i_C \quad \text{where } v_\mathrm{o} \neq 0 \;. \tag{5.14}$$

With this arrangement, the monitoring of instantaneous i_r and v_o allows information of the instantaneous r_L to be known. By absorbing this information into the control scheme, an adaptive feedback SM voltage controller, which basically varies α according to r_L, is obtained. Figure 5.8 gives an illustration how such an adaptive feedback control scheme can be implemented for the SMVC buck converter. Note also that $v_\mathrm{o} \neq 0$, which otherwise causes a division by zero problem. For analog implementation of this equation, a divider that saturates the computation signal at zero division may be used.

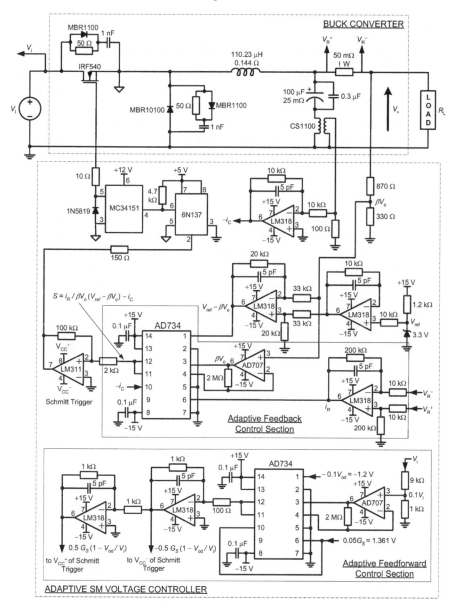

FIGURE 5.9
Full schematic diagram of the adaptive SMVC buck converter prototype.

5.5   Experimental Results and Discussions

This section gives a discussion on the results which are obtained with the adaptive feedforward and adaptive feedback control strategies. The experimental prototype has been constructed with specifications as shown in Table 4.1. Figure 5.9 shows the full schematic diagram of the proposed converter and controller. Separate tests are performed to evaluate the performance of the adaptive controllers with respect to line regulation and load regulation.

5.5.1   Line Variation

Figures 5.10(a)–5.10(d) show the experimental waveforms of the converter system at minimum and maximum input voltages, for the SM controller with and without using the adaptive feedforward control scheme. It can be seen that for both the cases $v_i = 18$ V and $v_i = 30$ V, with the same input voltage, the system with the adaptive feedforward control has switching period much closer to the desired switching period $T = 5$ μs. A plot of the measured average switching frequency versus different input voltages is shown in Fig. 5.11. The experimental data confirms more conclusively the capability of the adaptive feedforward control scheme in reducing the variation of $\overline{f_S}$. For the input voltage range 18 V $\leq v_i \leq$ 30 V, the frequency variation is reduced from $\pm 28.8\%$ to within $\pm 10.0\%$ of f_{Sd}. The improvement is less than the one obtained from simulation as shown in Fig. 5.5, which reduces the variation for the same input voltage range from $\pm 35\%$ to within $\pm 5\%$ of f_{Sd}. The deterioration is caused by practical components' variations and delay times. These factors were not taken into consideration in the modeling of the simulation program.

Figures 5.12(a) and 5.12(b) show the experimental waveforms of the converter system to which a step change of input voltage from minimum to maximum is applied. For the converter without the adaptive feedforward control, the enlarged view of the output voltage indicates that there is an upward DC shift when the input voltage steps up to a higher value. As explained earlier, this is due to the change in switching frequency, which increases or decreases the magnitude of the steady-state output voltage error caused by the imperfect feedback loop. Hence, for the converter with the adaptive feedforward control where the switching frequency variation is much reduced, there is less DC shift in the output voltage associated with the step change. This situation is experimentally captured and illustrated in Fig. 5.13. For the input voltage range 18 V $\leq v_i \leq$ 30 V, the line regulation is improved from 1.59% of $V_{o(v_i=24\ V)}$ to 0.17% of $V_{o(v_i=24\ V)}$ through the adaptive feedforward control scheme.

Figures 5.14(a) and 5.14(b) show the output voltage ripple waveforms of the converter operating at nominal load when v_i is sinusoidally varied from

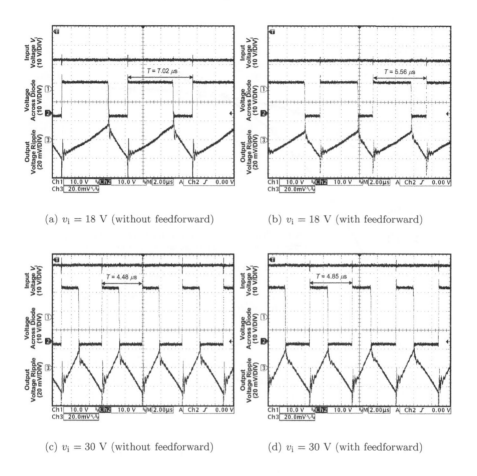

(a) $v_i = 18$ V (without feedforward)   (b) $v_i = 18$ V (with feedforward)

(c) $v_i = 30$ V (without feedforward)   (d) $v_i = 30$ V (with feedforward)

FIGURE 5.10

Experimental waveforms of input voltage v_i, voltage across freewheeling diode D, and output voltage ripple $\tilde{v}_o$ of the SMVC converter that is operating at input voltage (a) $v_i = 18$ V (without adaptive feedforward control); (b) $v_i = 18$ V (with adaptive feedforward control); (c) $v_i = 30$ V (without adaptive feedforward control); and (d) $v_i = 30$ V (with adaptive feedforward control), under nominal load resistance $r_L = 6$ Ω.

20.9 V to 27.1 V at a frequency of 100 Hz. The aim is to test the robustness of the converter against a slowly varying input voltage. Without the adaptive feedforward control, the maximum peak-to-peak output voltage is around 80 mV, i.e., the input voltage ripple rejection is -37.81 dB at 100 Hz. With the adaptive feedforward control, the maximum peak-to-peak output voltage is around 200 mV, i.e., the input voltage ripple rejection is -29.85 dB at 100 Hz. The slight deterioration in the audio-susceptibility performance il-

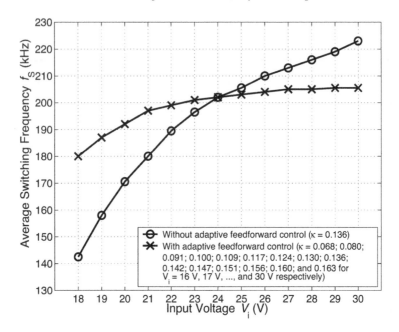

FIGURE 5.11

Experimentally measured average switching frequency $\overline{f_S}$ of the SMVC buck converter, with and without the incorporation of the adaptive feedforward control scheme, operating under line variation with input voltage $18\text{ V} \leq v_i \leq 30\text{ V}$ and at $r_L = 6\ \Omega$.

lustrates the main tradeoff in using the adaptive feedforward control scheme. Yet, in both cases, the converter still has an adequate audio-susceptibility performance.

It is also interesting to see that the output voltage varies twice as fast as the line variation. Without the adaptive feedforward control, the output voltage varies at the same frequency as the input voltage, i.e., 100 Hz. However, with the adaptive feedforward control, the output voltage varies at twice the frequency of the input voltage, i.e., 200 Hz. This can be explained by inspecting (4.27): $f_S = \frac{v_o\left(1 - \frac{v_o}{v_i}\right)}{2\kappa L}$. Without adaptive control, since κ is kept constant and f_S is varying with the change of input voltage v_i, both f_S and v_o vary at the same frequency as v_i. With the adaptive control, κ varies sinusoidally at 100 Hz. Since the adaptive control cannot fully eliminate the frequency variation due to line variation, f_S will also vary sinusoidally at 100 Hz. Hence, v_o will vary according to the product of κ and f_S, i.e., $\sin^2 2\pi100t \Rightarrow \cos 2\pi200t$.

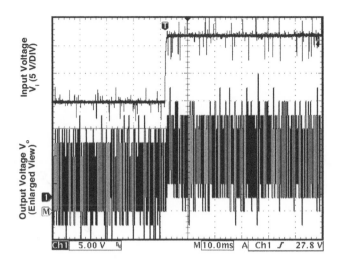

(a) Without adaptive feedforward control

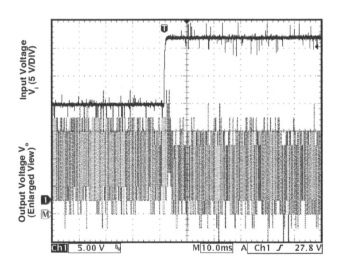

(b) With adaptive feedforward control

FIGURE 5.12
Experimental waveforms of input voltage v_i and enlarged view of output voltage v_o of the SMVC buck converter (a) without adaptive feedforward control and (b) with adaptive feedforward control, operating at step input voltage change from $v_i = 18$ V to $v_i = 30$ V, and at $r_L = 6\ \Omega$.

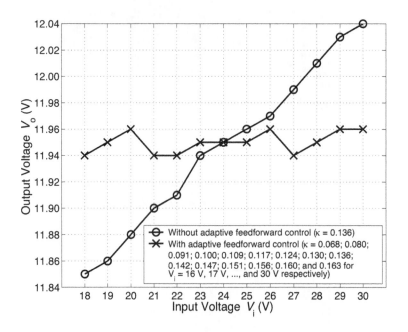

FIGURE 5.13

Experimentally measured output voltage v_o of the SMVC buck converter, with and without the incorporation of the adaptive feedforward control scheme, operating under line variation with input voltage $18 \text{ V} \leq v_i \leq 30 \text{ V}$ and at $r_L = 6 \ \Omega$.

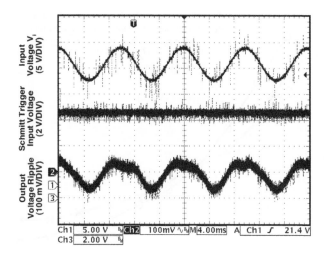

(a) Without adaptive feedforward control

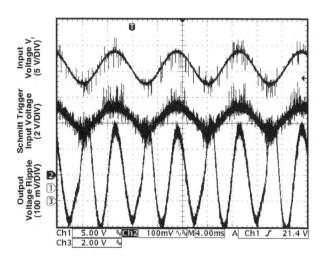

(b) With adaptive feedforward control

FIGURE 5.14
Experimental waveforms of input voltage v_i, positive input voltage to the Schmitt trigger V_{CC}^+, and output voltage ripple $\tilde{v}_o$ of the SMVC buck converter (a) without adaptive feedforward control and (b) with adaptive feedforward control, operating at sinusoidally varying input voltage from $v_i = 20.9$ V to $v_i = 27.1$ V, and at $r_L = 6$ Ω.

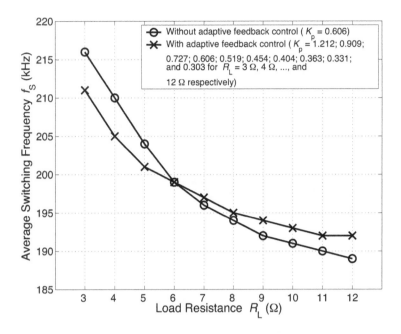

FIGURE 5.15
Experimentally measured average switching frequency $\overline{f}_S$ of the SMVC buck converter, with and without the incorporation of the adaptive feedback control scheme, operating under load variation with load resistance $3\ \Omega \leq r_L \leq 12\ \Omega$ and at $v_i = 24$ V.

5.5.2 Load Variation

Figures 5.15 and 5.16 show the experimental data of the converter system at different load resistances for the SM controller with and without the adaptive feedback control scheme. From Fig. 5.15, it can be seen that with the adaptive feedback control, the variation of the switching frequency with respect to the load resistance improves from an average of $\frac{d\overline{f_S}}{dr_L} = -3.0$ kHz/Ω (without adaptive feedback control) to an average of $\frac{d\overline{f_S}}{dr_L} = -2.1$ kHz/Ω (with adaptive feedback control). Thus, for the load resistance range $3\ \Omega \leq r_L \leq 12\ \Omega$, the frequency variation has been reduced from $\pm 8.5\%$ to within $\pm 6.0\%$ of f_{Sd}. This verifies the capability of the adaptive feedback control scheme in suppressing the variation of switching frequency caused by load variation. Furthermore, it should be pointed out that the quantitative difference between the results obtained from the experiment and the simulation in Fig. 5.7 is due to practical components' variations and delay times that were not modeled in the simulation program.

The reduction in the switching frequency variation with the adaptive feedback control is predictably the outcome of better performance in the load

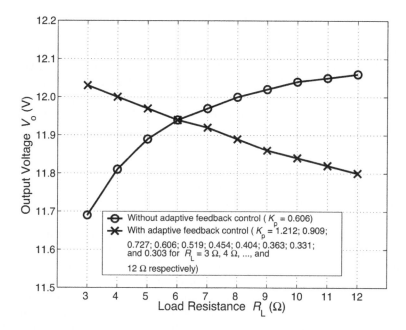

FIGURE 5.16
Experimentally measured output voltage v_o of the SMVC buck converter, with and without the incorporation of the adaptive feedback control scheme, operating under load variation with load resistance $3\ \Omega \leq r_L \leq 12\ \Omega$ and at $v_i = 24$ V.

regulation. This has been explained earlier and can be observed from Fig. 5.16. Without the adaptive feedback control, there is a 0.37 V deviation (i.e., 3.1% of $V_{o(\text{nominal load})}$) in v_o for the entire load range, i.e., load regulation $\frac{dv_o}{dr_L}$ averages at 0.040 V/Ω. With the adaptive feedback control, there is a -0.23 V deviation (i.e., -1.9% of $V_{o(\text{nominal load})}$) in v_o for the entire load range, i.e., load regulation $\frac{dv_o}{dr_L}$ averages at -0.026 V/Ω. Noticeably, the incorporation of the adaptive feedback control scheme has changed the way in which the output voltage is regulated. This is evident from the fact that the output voltage without the adaptive feedback control increases with increasing resistance whereas the output voltage with the adaptive feedback control decreases with increasing resistance. Recall that α is varied as the load changes. Since α is effectively the DC gain parameter in the controller, it is actually controlling the steady-state output voltage. Thus, when the load resistance increases, the output voltage of the converter (without adaptive feedback control) should supposedly increase, and its switching frequency should drop below the nominal value. However, when α is adaptive (with the adaptive feedback control) and could react to the increment in the load resistance, it actually decreases, causing the output voltage to decrease. The switching frequency thus increases accordingly.

Figures 5.17(a) and 5.17(b) show the output waveforms of the operation with load resistance that alternates between $r_L = 12\ \Omega$ and $r_L = 3\ \Omega$. The comparison shows that the incorporation of the adaptive feedback control scheme will have little effect on the transient performance. Specifically, the overshoot voltage ripple is increased from 120 mV (without adaptive feedback) to 180 mV (with adaptive feedback), and the setting time from 60 ms to 70 ms.

Remarks

From these case studies, the ability of the adaptive feedback control scheme to reduce the switching frequency variation is seemingly insignificant as compared to the adaptive feedforward control scheme. However, recalling that the switching frequency variations are much larger for larger load variations, the ability of the adaptive feedback control scheme to suppress variations in such circumstances will be more easily appreciated. Moreover, the inclusion of the adaptive feedback control scheme also constitutes a significant improvement in the load regulation of the converter, which is also an important aspect of its application. Yet, considering that the employment of the scheme requires additional circuitries and a current sensor, which adds more complexity to the SM controller, the decision for its adoption in converter control is therefore application specific.

Furthermore, the implementation of this kind of adaptive control schemes is generally much easier than the implementation of the adaptive control schemes in conventional linear controllers. This is because in the SM control, the relationship of the control equation is directly corresponding to the input voltage and the load resistance. Hence, the implementation of the adaptive control schemes can be realized by directly monitoring these operating conditions. However, for the conventional linear controllers, the gain parameters that give the optimal control normally require some form of indirect computation to generate. Its realization in analog form is therefore far more complex. Finally, our findings clearly suggest that the SM controller would be a convenient platform for the implementation of adaptive control that is required for some systems.

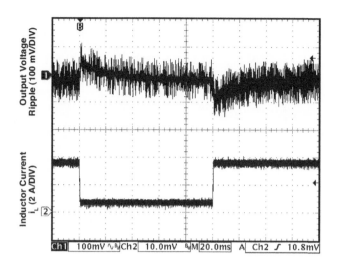

(a) Without adaptive feedback control

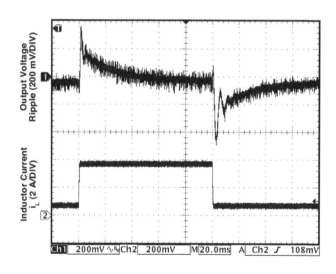

(b) With adaptive feedback control

FIGURE 5.17
Experimental waveforms of output voltage ripple $\tilde{v}_o$ and inductor current i_L of the SMVC buck converter (a) without adaptive feedback control and (b) with adaptive feedback control, operating at $v_i = 24$ V and step load changes that alternates between $r_L = 12\ \Omega$ and $r_L = 3\ \Omega$.

6

General Approach of Deriving PWM-Based Sliding Mode Controller for Power Converters in Continuous Conduction Mode

CONTENTS

6.1 Introduction

In the previous chapters, we have discussed the use of a hysteresis band in the switching relay to suppress the switching frequency variation, and the introduction of the adaptive control schemes to maintain the switching frequency within a tolerance range against the line or load variations. In this and the following two chapters, we describe an effective and practical approach for controlling the switching frequency of the SM controller via fixed-frequency pulse-width modulation (PWM).

Our discussion will be organized in three chapters, and will address the various issues concerning the development of the so-called fixed-frequency PWM-based SM controllers. A complete exposition of the design procedure and its application to converters will be presented.

In this chapter, we derive the PWM-based SM voltage controller for DC–DC converters operating in continuous conduction mode (CCM). Our aim is to present simple and ready-to-use control equations for implementation of the PWM-based SM voltage controller. Although the discussion covers only the buck, boost, and buck-boost converters, the approach is applicable to all other DC–DC converter types.

In the next chapter, we will derive the PWM-based SM controller for DC–DC converters operating in discontinuous conduction mode. Finally, in Chapter 8, we will give a detailed discussion on the practical design and implementation issues of the PWM-based SM controllers.

6.2 Background

The technique of pulse-width modulation consists in comparing a desired analog control signal v_c with a ramp signal, from which a pulse-like output switching signal having the same frequency as the ramp signal will be generated. The advantage is that by fixing the frequency of the ramp, the frequency of the output switching signal will be constant. Thus, by employing this modulation technique in SM control, a fixed-frequency PWM-based SM controller can be obtained. Meanwhile, it should be stressed that the application of the PWM technique to SM control does not contradict the basic principle of classical PWM controllers used in power electronics. The difference is the way in which the control signal v_c is formulated. SM controllers are based on the SM control law whereas classical PWM controllers are based on the linear control law. Hence, hereon, the term PWM-based SM controller essentially refers to a pulse-width modulator that employs an equivalent control (derived by applying the SM control technique) to generate a control signal to be compared with the fixed-frequency ramp in the modulator.

To implement such a controller, a relationship between SM control and duty-ratio control is required. The idea can be rooted back to one of the earliest papers on SM-controlled DC–DC converters [104], which suggests that under SM control operation, the control signal of *equivalent control approach* u_{eq} in SM control is equivalent to the *duty ratio* d of a PWM controller. However, the theoretical framework supporting such a claim was absent from the paper. It was some time later when Sira-Ramirez *et al.* [78] proposed a geometric framework to map the PWM feedback control onto SM control that the proof was rigorously shown in a companion paper [79]. Specifically, what has been shown is that as the switching frequency tends to infinity, the averaged dynamics of an SM-controlled system is equivalent to the averaged dynamics of a PWM-controlled system, thus establishing the relationship $u_{\text{eq}} = d$. Independently, the same correlation was derived in Martinez *et al.* [48], where the nonlinear PWM continuous control was compared with an equivalent control. In their method, an average representation of the converter model was employed. Hence, the migration of an SM controller from being HM-based to PWM-based is made possible. Unfortunately, the theory was not exploited to expedite the development of such controllers.

The first useful clue as to how the PWM technique can be applied to develop fixed-frequency SM controllers is probably due to Nguyen and Lee [63]. In two other related works [44, 45], the *state-space averaging technique* was incorporated into the controller's modeling. By doing so, PWM duty-ratio control can be directly applied to the implementation of SM controllers. For a deeper understanding of the topic, interested readers may refer to [44, 45, 63, 78, 79, 104].

6.3   The Approach

In the following, the modeling method and the detailed procedures for designing SM controllers for DC–DC converters in CCM operation will be provided.

6.3.1   System Modeling

The first step to the design of an SM controller is to develop a state-space description of the converter model in terms of the desired control variables (i.e., voltage and/or current). Our focus in this chapter is the application of SM control to converters operating in CCM. The controller under study is a second-order proportional-integral-derivative (PID) SM voltage-mode controller. Unlike most previously proposed SM voltage controllers, it takes into account an additional voltage error integral term in the control computation to reduce the steady-state error of the practical SM-controlled system. This is commonly known as *integral SM control* [102] and its application in power converters has captured some recent interests [23, 55, 84].

Figure 6.1 shows the schematic diagrams of the three PID sliding mode voltage controlled (SMVC) DC–DC converters in the conventional HM configuration. Here, C, L, and r_{L} denote the capacitance, inductance, and instantaneous load resistance of the converters, respectively; i_C, i_L, and i_r denote

Sliding Mode Control of Switching Power Converters

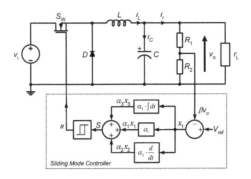

(a) SMVC buck converter

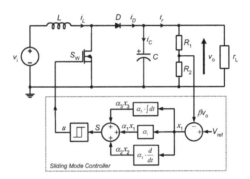

(b) SMVC boost converter

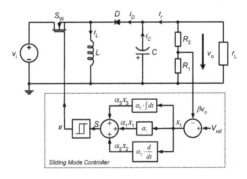

(c) SMVC buck-boost converter

FIGURE 6.1
Schematic diagrams of conventional HM-based PID SMVC converters.

the instantaneous capacitor, inductor, and load currents, respectively; V_{ref}, v_i, and βv_o denote the reference, instantaneous input, and instantaneous output voltages, respectively; β denotes the feedback network ratio; and $u = 0$ or 1 is the switching state of power switch S_W.

In the case of PID SMVC converters, the control variable x may be expressed in the following form:

$$
x = \begin{bmatrix} x_1 \\ x_2 \\ x_3 \end{bmatrix} = \begin{bmatrix} V_{ref} - \beta v_o \\ \dfrac{d(V_{ref} - \beta v_o)}{dt} \\ \int (V_{ref} - \beta v_o)dt \end{bmatrix} \tag{6.1}
$$

where x_1, x_2, and x_3 are the *voltage error*, the *voltage error dynamics* (or the rate of change of voltage error), and the *integral of voltage error*, respectively.

Substitution of the converters' behavioral models under CCM into (6.1) produces the following control variable descriptions: x_{buck}, x_{boost}, and $x_{buck-boost}$ for buck, boost, and buck-boost converter, respectively.

$$
x_{buck} = \begin{bmatrix} x_1 & = & V_{ref} - \beta v_o \\ x_2 & = & \dfrac{\beta v_o}{r_L C} + \displaystyle\int \dfrac{\beta(v_o - v_i u)}{LC}dt \\ x_3 & = & \int x_1 dt \end{bmatrix} ; \tag{6.2}
$$

$$
x_{boost} = \begin{bmatrix} x_1 & = & V_{ref} - \beta v_o \\ x_2 & = & \dfrac{\beta v_o}{r_L C} + \displaystyle\int \dfrac{\beta(v_o - v_i)\bar u}{LC}dt \\ x_3 & = & \int x_1 dt \end{bmatrix} ; \tag{6.3}
$$

$$
x_{buck-boost} = \begin{bmatrix} x_1 & = & V_{ref} - \beta v_o \\ x_2 & = & \dfrac{\beta v_o}{r_L C} + \displaystyle\int \dfrac{\beta v_o \bar u}{LC}dt \\ x_3 & = & \int x_1 dt \end{bmatrix} . \tag{6.4}
$$

Next, the time differentiation of equations (6.2), (6.3), and (6.4) produce the state-space descriptions required for the controller design of the respective converters.

For buck converter:

$$
\begin{bmatrix} \dot x_1 \\ \dot x_2 \\ \dot x_3 \end{bmatrix} = \begin{bmatrix} 0 & 1 & 0 \\ 0 & -\dfrac{1}{r_L C} & 0 \\ 1 & 0 & 0 \end{bmatrix} \begin{bmatrix} x_1 \\ x_2 \\ x_3 \end{bmatrix} + \begin{bmatrix} 0 \\ -\dfrac{\beta v_i}{LC} \\ 0 \end{bmatrix} u + \begin{bmatrix} 0 \\ \dfrac{\beta v_o}{LC} \\ 0 \end{bmatrix} . \tag{6.5}
$$

For boost converter:

$$
\begin{bmatrix} \dot x_1 \\ \dot x_2 \\ \dot x_3 \end{bmatrix} = \begin{bmatrix} 0 & 1 & 0 \\ 0 & -\dfrac{1}{r_L C} & 0 \\ 1 & 0 & 0 \end{bmatrix} \begin{bmatrix} x_1 \\ x_2 \\ x_3 \end{bmatrix} + \begin{bmatrix} 0 \\ \dfrac{\beta v_o}{LC} - \dfrac{\beta v_i}{LC} \\ 0 \end{bmatrix} \bar u . \tag{6.6}
$$

TABLE 6.1
Descriptions of SMVC buck, boost, and buck-boost converters operating in CCM.

Type of Converter	**A**	**B**	**D**	v
Buck	$\begin{bmatrix} 0 & 1 & 0 \\ 0 & -\frac{1}{r_L C} & 0 \\ 1 & 0 & 0 \end{bmatrix}$	$\begin{bmatrix} 0 \\ -\frac{\beta v_i}{LC} \\ 0 \end{bmatrix}$	$\begin{bmatrix} 0 \\ \frac{\beta v_o}{LC} \\ 0 \end{bmatrix}$	u
Boost	$\begin{bmatrix} 0 & 1 & 0 \\ 0 & -\frac{1}{r_L C} & 0 \\ 1 & 0 & 0 \end{bmatrix}$	$\begin{bmatrix} 0 \\ \frac{\beta v_o}{LC} - \frac{\beta v_i}{LC} \\ 0 \end{bmatrix}$	$\begin{bmatrix} 0 \\ 0 \\ 0 \end{bmatrix}$	$\bar{u}$
Buck-Boost	$\begin{bmatrix} 0 & 1 & 0 \\ 0 & -\frac{1}{r_L C} & 0 \\ 1 & 0 & 0 \end{bmatrix}$	$\begin{bmatrix} 0 \\ \frac{\beta v_o}{LC} \\ 0 \end{bmatrix}$	$\begin{bmatrix} 0 \\ 0 \\ 0 \end{bmatrix}$	$\bar{u}$

For buck-boost:

$$\begin{bmatrix} \dot{x}_1 \\ \dot{x}_2 \\ \dot{x}_3 \end{bmatrix} = \begin{bmatrix} 0 & 1 & 0 \\ 0 & -\frac{1}{r_L C} & 0 \\ 1 & 0 & 0 \end{bmatrix} \begin{bmatrix} x_1 \\ x_2 \\ x_3 \end{bmatrix} + \begin{bmatrix} 0 \\ \frac{\beta v_o}{LC} \\ 0 \end{bmatrix} \bar{u} \qquad (6.7)$$

where $\bar{u} = 1 - u$ is the inverse logic of u, and is used particularly for modeling the boost and buck-boost topologies. Rearrangement of the state-space descriptions (6.5), (6.6), and (6.7) into the standard form gives

$$\dot{x} = \mathbf{A}x + \mathbf{B}v + \mathbf{D} \qquad (6.8)$$

where $v = u$ or $\bar{u}$ (depending on topology). Results are summarized in the tabulated format shown in Table 6.1.

6.3.2 Controller Design

Having obtained the state-space descriptions, the next stage is the design of the controller. For these systems, it is appropriate to have a general SM control law that adopts a switching function such as

$$u = \begin{cases} 1 & \text{when } S > 0 \\ 0 & \text{when } S < 0 \end{cases} \qquad (6.9)$$

where S is the instantaneous state trajectory, and is described as

$$S = \alpha_1 x_1 + \alpha_2 x_2 + \alpha_3 x_3 = \boldsymbol{J}^{\mathrm{T}} \boldsymbol{x}, \tag{6.10}$$

with $\boldsymbol{J}^{\mathrm{T}} = [\alpha_1 \ \alpha_2 \ \alpha_3]$ and $\alpha_1, \alpha_2,$ and α_3 representing the sliding coefficients.

6.3.2.1 Derivation of Existence Conditions

Next, let us consider how the existence conditions of SM control operation are obtained for the converters. Since the fixed-frequency PWM-based controller is a translated form of the HM-based controller, the model for the latter must first be derived. Therefore, the discussion in this section is applicable for both the PWM-based and HM-based controllers.

To ensure the existence of SM operation, the local reachability condition

$$\lim_{S \to 0} S \cdot \dot{S} < 0 \tag{6.11}$$

must be satisfied. This can be expressed as

$$\begin{cases} \dot{S}_{S \to 0^+} = \boldsymbol{J}^{\mathrm{T}} \boldsymbol{A} \boldsymbol{x} + \boldsymbol{J}^{\mathrm{T}} \boldsymbol{B} v_{S \to 0^+} + \boldsymbol{J}^{\mathrm{T}} \boldsymbol{D} < 0 \\ \dot{S}_{S \to 0^-} = \boldsymbol{J}^{\mathrm{T}} \boldsymbol{A} \boldsymbol{x} + \boldsymbol{J}^{\mathrm{T}} \boldsymbol{B} v_{S \to 0^-} + \boldsymbol{J}^{\mathrm{T}} \boldsymbol{D} > 0 \end{cases}. \tag{6.12}$$

Illustrations are provided for the buck and the boost converters.

Example—Buck Converter

- Case 1: $S \to 0^+$, $\dot{S} < 0$:
  Substitution of $v_{S \to 0^+} = u = 1$ and the matrices in Table 6.1 into (6.12) gives

$$-\alpha_1 \frac{\beta i_C}{C} + \alpha_2 \frac{\beta i_C}{r_{\mathrm{L}} C^2} + \alpha_3 (V_{\mathrm{ref}} - \beta v_{\mathrm{o}}) - \alpha_2 \frac{\beta v_{\mathrm{i}}}{LC} + \alpha_2 \frac{\beta v_{\mathrm{o}}}{LC} < 0. \tag{6.13}$$

- Case 2: $S \to 0^-$, $\dot{S} > 0$:
  Substitution of $v_{S \to 0^-} = u = 0$ and the matrices in Table 6.1 into (6.12) gives

$$-\alpha_1 \frac{\beta i_C}{C} + \alpha_2 \frac{\beta i_C}{r_{\mathrm{L}} C^2} + \alpha_3 (V_{\mathrm{ref}} - \beta v_{\mathrm{o}}) + \alpha_2 \frac{\beta v_{\mathrm{o}}}{LC} > 0. \tag{6.14}$$

Finally, the combination of (6.13) and (6.14) gives the simplified existence condition

$$0 < -\beta L \left(\frac{\alpha_1}{\alpha_2} - \frac{1}{r_{\mathrm{L}} C} \right) i_C + LC \frac{\alpha_3}{\alpha_2} (V_{\mathrm{ref}} - \beta V_{\mathrm{o}}) + \beta v_{\mathrm{o}} < \beta v_{\mathrm{i}}. \tag{6.15}$$

TABLE 6.2
Existence conditions of buck, boost, and buck-boost converters operating in CCM.

Buck	$0 < -\beta L \left(\frac{\alpha_1}{\alpha_2} - \frac{1}{r_L C} \right) i_C + LC \frac{\alpha_3}{\alpha_2} (V_{\text{ref}} - \beta v_o) + \beta v_o < \beta v_i$
Boost	$0 < \beta L \left(\frac{\alpha_1}{\alpha_2} - \frac{1}{r_L C} \right) i_C - LC \frac{\alpha_3}{\alpha_2} (V_{\text{ref}} - \beta v_o) < \beta(v_o - v_i)$
Buck-Boost	$0 < \beta L \left(\frac{\alpha_1}{\alpha_2} - \frac{1}{r_L C} \right) i_C - LC \frac{\alpha_3}{\alpha_2} (V_{\text{ref}} - \beta v_o) < \beta v_o$

Example—Boost Converter

- Case 1: $S \to 0^+$, $\dot{S} < 0$:
  Substitution of $v_{S \to 0^+} = \bar{u} = 0$ and the matrices in Table 6.1 into (6.12) gives

$$-\alpha_1 \frac{\beta i_C}{C} + \alpha_2 \frac{\beta i_C}{r_L C^2} + \alpha_3 (V_{\text{ref}} - \beta v_o) < 0. \tag{6.16}$$

- Case 2: $S \to 0^-$, $\dot{S} > 0$:
  Substitution of $v_{S \to 0^-} = \bar{u} = 1$ and the matrices in Table 6.1 into (6.12) gives

$$-\alpha_1 \frac{\beta i_C}{C} + \alpha_2 \frac{\beta i_C}{r_L C^2} + \alpha_3 (V_{\text{ref}} - \beta v_o) - \alpha_2 \frac{\beta v_i}{LC} + \alpha_2 \frac{\beta v_o}{LC} > 0. \tag{6.17}$$

Finally, the combination of (6.16) and (6.17) gives the simplified existence condition

$$0 < \beta L \left(\frac{\alpha_1}{\alpha_2} - \frac{1}{r_L C} \right) i_C - LC \frac{\alpha_3}{\alpha_2} (V_{\text{ref}} - \beta v_o) < \beta(v_o - v_i). \tag{6.18}$$

The derived existence conditions for the buck, boost, and buck-boost converters are tabulated in Table 6.2. The selection of sliding coefficients for the controller of each converter must comply to its stated inequalities. It is important to ensure that the circuit component tolerances and the complete ranges of operating conditions are taken into account when evaluating the inequalities. This assures the compliance of the existence condition for the full operating ranges of the converters.

6.3.2.2 Derivation of Control Equations for PWM-Based Controller

The conventional SM controller implementation based on HM requires only control equations (6.9) and (6.10). The same controller is used for all three converters. However, if the PWM-based SM voltage controller is to be adopted, an indirect translation of the SM control law is required so that pulse-width modulation can be used in lieu of hysteretic modulation (see Fig. 3.3). This results in a unique controller for each converter since their control equations are all different. The derivation process of this controller can be summarized in two steps. Firstly, the equivalent control signal u_{eq}, which is a smooth function of the discrete input function u, is formulated using the *invariance conditions* by setting the time differentiation of (6.10) as $\dot{S} = 0$ [102]. Secondly, the equivalent control function is translated to the instantaneous duty ratio d of the pulse-width modulator [104]. Illustrations are provided for the buck and the boost converters.

Example—Buck Converter

Equating $\dot{S} = \mathbf{J}^{\mathrm{T}}\mathbf{A}\mathbf{x} + \mathbf{J}^{\mathrm{T}}\mathbf{B}u_{eq} + \mathbf{D} = 0$ yields the equivalent control function

$$u_{eq} = -\left[\mathbf{J}^{\mathrm{T}}\mathbf{B}\right]^{-1}\mathbf{J}^{\mathrm{T}}\left[\mathbf{A}\mathbf{x} + \mathbf{D}\right] \tag{6.19}$$

$$= -\frac{\beta L}{\beta v_i}\left(\frac{\alpha_1}{\alpha_2} - \frac{1}{r_L C}\right)i_C + \frac{\alpha_3 LC}{\alpha_2 \beta v_i}\left(V_{ref} - \beta v_o\right) + \frac{v_o}{v_i}$$

where u_{eq} is continuous and $0 < u_{eq} < 1$. Substitution of (6.19) into the inequality gives

$$0 < u_{eq} = -\frac{\beta L}{\beta v_i}\left(\frac{\alpha_1}{\alpha_2} - \frac{1}{r_L C}\right)i_C + \frac{\alpha_3 LC}{\alpha_2 \beta v_i}\left(V_{ref} - \beta v_o\right) + \frac{v_o}{v_i} < 1. \tag{6.20}$$

Multiplication of the inequality by βv_i gives

$$0 < u_{eq}{}^* = -\beta L\left(\frac{\alpha_1}{\alpha_2} - \frac{1}{r_L C}\right)i_C + LC\frac{\alpha_3}{\alpha_2}\left(V_{ref} - \beta v_o\right) + \beta v_o < \beta v_i. \tag{6.21}$$

Finally, the translation of the equivalent control function (6.21) to the duty ratio d, where $0 < d = \frac{v_c}{\hat{v}_{ramp}} < 1$, gives the following relationships for the control signal v_c and ramp signal $\hat{v}_{ramp}$ for the practical implementation of the PWM-based SM controller:

$$v_c = u_{eq}{}^* = -\beta L\left(\frac{\alpha_1}{\alpha_2} - \frac{1}{r_L C}\right)i_C + \frac{\alpha_3}{\alpha_2}LC\left(V_{ref} - \beta v_o\right) + \beta v_o \tag{6.22}$$

and

$$\hat{v}_{ramp} = \beta v_i. \tag{6.23}$$

Example—Boost Converter

Equating $\dot{S} = \boldsymbol{J}^{\mathrm{T}}\boldsymbol{A}\boldsymbol{x} + \boldsymbol{J}^{\mathrm{T}}\boldsymbol{B}\bar{u}_{\mathrm{eq}} = 0$ yields the equivalent control function

$$
\begin{aligned}
\bar{u}_{\mathrm{eq}} &= -\left[\boldsymbol{J}^{\mathrm{T}}\boldsymbol{B}\right]^{-1}\boldsymbol{J}^{\mathrm{T}}\boldsymbol{A}\boldsymbol{x} \tag{6.24}\\
&= \frac{\beta L}{\beta\left(v_{\mathrm{o}} - v_{\mathrm{i}}\right)}\left(\frac{\alpha_1}{\alpha_2} - \frac{1}{r_{\mathrm{L}}C}\right)i_C - \frac{\alpha_3 LC}{\alpha_2\beta\left(v_{\mathrm{o}} - v_{\mathrm{i}}\right)}\left(V_{\mathrm{ref}} - \beta v_{\mathrm{o}}\right)
\end{aligned}
$$

where $\bar{u}_{\mathrm{eq}}$ is continuous and $0 < \bar{u}_{\mathrm{eq}} < 1$. Substitution of (6.24) into the inequality gives

$$
\begin{aligned}
0 < \bar{u}_{\mathrm{eq}} = {} & \frac{\beta L}{\beta\left(v_{\mathrm{o}} - v_{\mathrm{i}}\right)}\left(\frac{\alpha_1}{\alpha_2} - \frac{1}{r_{\mathrm{L}}C}\right)i_C \tag{6.25}\\
& -\frac{\alpha_3 LC}{\alpha_2\beta\left(v_{\mathrm{o}} - v_{\mathrm{i}}\right)}\left(V_{\mathrm{ref}} - \beta v_{\mathrm{o}}\right) < 1.
\end{aligned}
$$

Since $u = 1 - \bar{u}$, which also implies $u_{\mathrm{eq}} = 1 - \bar{u}_{\mathrm{eq}}$, the inequality can be rewritten as

$$
\begin{aligned}
0 < u_{\mathrm{eq}} = {} & 1 - \frac{\beta L}{\beta\left(v_{\mathrm{o}} - v_{\mathrm{i}}\right)}\left(\frac{\alpha_1}{\alpha_2} - \frac{1}{r_{\mathrm{L}}C}\right)i_C \\
& +\frac{\alpha_3 LC}{\alpha_2\beta\left(v_{\mathrm{o}} - v_{\mathrm{i}}\right)}\left(V_{\mathrm{ref}} - \beta v_{\mathrm{o}}\right) < 1. \tag{6.26}
\end{aligned}
$$

Multiplication of the inequality by $\beta(v_{\mathrm{o}} - v_{\mathrm{i}})$ gives

$$
\begin{aligned}
0 < u_{\mathrm{eq}}{}^* = {} & -\beta L\left(\frac{\alpha_1}{\alpha_2} - \frac{1}{r_{\mathrm{L}}C}\right)i_C + LC\frac{\alpha_3}{\alpha_2}\left(V_{\mathrm{ref}} - \beta v_{\mathrm{o}}\right) \\
& +\beta\left(v_{\mathrm{o}} - v_{\mathrm{i}}\right) < \beta\left(v_{\mathrm{o}} - v_{\mathrm{i}}\right). \tag{6.27}
\end{aligned}
$$

Finally, the translation of the equivalent control function (6.27) to the duty ratio d, where $0 < d = \frac{v_c}{\hat{v}_{\mathrm{ramp}}} < 1$, gives the following relationships for the control signal v_c and ramp signal $\hat{v}_{\mathrm{ramp}}$ for the practical implementation of the PWM-based SM controller:

$$
v_{\mathrm{c}} = u_{\mathrm{eq}}{}^* = -\beta L\left(\frac{\alpha_1}{\alpha_2} - \frac{1}{r_{\mathrm{L}}C}\right)i_C + LC\frac{\alpha_3}{\alpha_2}\left(V_{\mathrm{ref}} - \beta v_{\mathrm{o}}\right) + \beta\left(v_{\mathrm{o}} - v_{\mathrm{i}}\right) \tag{6.28}
$$

and

$$
\hat{v}_{\mathrm{ramp}} = \beta\left(v_{\mathrm{o}} - v_{\mathrm{i}}\right). \tag{6.29}
$$

The control equations required for the implementation of the respective PWM-based SMVC converters are tabulated in Table 6.3. K_{p1} and K_{p2} are the constant gain parameters for the feedback signals i_C and $(V_{\mathrm{ref}} - \beta v_{\mathrm{o}})$. The values of K_{p1} and K_{p2} can be found in terms of converter's parameters L, C, and r_{L}, and the values of sliding coefficients α_1, α_2, and α_3, which must comply with the existence condition stated in Table 6.2.

TABLE 6.3

Control equations of PWM-based SMVC buck, boost, and buck-boost converters operating in CCM. K_{p1} and K_{p2} are calculated using $K_{p1} = \beta L \left(\frac{\alpha_1}{\alpha_2} - \frac{1}{r_L C} \right)$ and $K_{p2} = \frac{\alpha_3}{\alpha_2} LC$.

Type of Converter	v_c	$\hat{v}_{\text{ramp}}$
Buck	$-K_{p1} i_C + K_{p2} \left(V_{\text{ref}} - \beta v_o \right) + \beta v_o$	βv_i
Boost	$-K_{p1} i_C + K_{p2} \left(V_{\text{ref}} - \beta v_o \right) + \beta \left(v_o - v_i \right)$	$\beta \left(v_o - v_i \right)$
Buck-Boost	$-K_{p1} i_C + K_{p2} \left(V_{\text{ref}} - \beta v_o \right) + \beta v_o$	βv_o

6.3.3 Remarks

6.3.3.1 Controller Structure

Figure 6.2 shows the schematic diagrams of the respective PWM-based SMVC converters. The controller design is based on the equations illustrated in Table 6.3. Careful examination of these circuits reveal that they basically adopt the same structure as the PWM proportional-derivative (PD) linear control, but with additional components consisting of the instantaneous input voltage βv_i and/or the instantaneous output voltage βv_o. These are the components contributing to the nonlinearity of the feedback control, and are therefore the key properties keeping the controller robust to load and line regulations under wide operating ranges. It should also be noted that the integral term of the control variable, i.e., x_3, is implicitly hidden in the control variable v_o. In case of a large disturbance, this component is highly influential in the control. However, when the steady state is reached, v_o actually becomes a fixed point, thereby destroying the integral control. The equation then reduces to the PWM PD linear controller form.

6.3.3.2 Performance Comparison with HM-Based SM Controllers

It should be noted that similar to the HM-based SM controller, the PWM-based controller is not absolutely robust to line and load variations. As discussed, its robustness improves with switching frequency and full robustness can only be achieved when the switching frequency is infinite. Now, considering that both line and load variations will vary the switching frequency and hence deteriorate the DC regulation in the conventional HM-based SM-controlled converters, the adoption of the constant frequency PWM technique

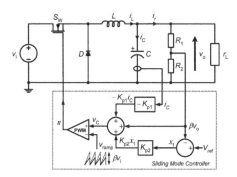

(a) SMVC buck converter

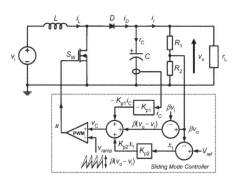

(b) SMVC boost converter

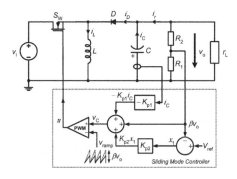

(c) SMVC buck-boost converter

FIGURE 6.2
Schematic diagrams of the PWM-based PID SMVC converters.

in the SM controllers will provide comparatively better steady-state line and load regulated converters. On the other hand, the dynamic responses of the two controllers (assuming both types of controllers to operate at a similar averaged constant switching frequency) will not differ significantly. Recall that in an SM-controlled system, the dynamic behavior is mainly determined by the sliding coefficients. Apparently, if both types of controllers have the same set of sliding coefficients and are operating in the same frequency range, their dynamic responses will be similar.

6.3.3.3 Comparing the PWM-Based SM Controller Approach to the Nonlinear PWM Controller Design Approach

It is also interesting to note that the PWM-based SM controllers can be viewed as a type of nonlinear state-feedback controllers designed from nonlinear "per-cycle averaged models" of the converters. In fact, the method reported in [48] can be used to derive the same nonlinear duty-ratio expression as provided in this chapter. The method used in nonlinear PWM controller design is to assume a large-signal average model of the converter at the start of the derivation [48]. However, for the PWM-based SM controller, the model of the converter is retained in its discrete state-space form throughout the derivation. When the duty ratio is equivalent to the equivalent control, i.e., $d = u_{eq}$, an averaging process takes place. Note that u_{eq} describes the *instantaneous* control required for SM control operation to occur. Hence, when the duty-ratio control is employed to implement the u_{eq} over a finite time period of a switching cycle, the required instantaneity is spread over a switching cycle. Therefore, this can be considered as "per-cycle averaging" of the converter model. Thus, a main difference of our approach from that of nonlinear PWM control is that while the same assumption of an "averaged model" holds, our approach, which only averages the model during the implementation of the controller, retains much of the converter dynamics. This results in a set of design restrictions in the existence conditions, which are evolved from the instantaneous dynamics of the converter, as required by the SM control theory. Such design restrictions are absent from the nonlinear PWM controller design approach. Finally, knowing that the abidance of the existence condition implies that the system's trajectory will strictly follow the desired sliding surface, the system's dynamic response will strictly obey the designed dynamics. Such stringency is not present in the nonlinear PWM controller design approach.

6.3.3.4 Load Resistance Dependence

A close inspection of the control equations reveals that the control signal v_c is actually load dependent. Thus, for the controller to have good regulation against load changes, the instantaneous value of r_L should be fed back. However, this would require additional sensors and cumbersome computations. On the other hand, the dependence and sensitivity of v_c on the load can be minimized by a proper design of sliding coefficient such that $\frac{\alpha_1}{\alpha_2} \gg \frac{1}{r_L C}$. In such circumstances, the design value of load resistance can be made a constant parameter $R_{L(nom)}$. If this is adopted, the real system's dynamics during SM operation will be changed from the ideal case of

$$\frac{d^2x_1}{dt^2} + \frac{\alpha_1}{\alpha_2} \cdot \frac{dx_1}{dt} + \frac{\alpha_3}{\alpha_2} \cdot x_1 = 0 \tag{6.30}$$

to the actual case of

$$\frac{d^2x_1}{dt^2} + \left(\frac{\alpha_1}{\alpha_2} + \frac{1}{r_L(t)C} - \frac{1}{R_{L(nom)}(t)C}\right) \cdot \frac{dx_1}{dt} + \frac{\alpha_3}{\alpha_2} \cdot x_1 = 0, \tag{6.31}$$

where $r_L(t) \neq R_{L(nom)}$ is the instantaneous load resistance, when the load differs from the design value. However, since $\frac{\alpha_1}{\alpha_2} \gg \frac{1}{r_L C}$, the dynamics of the actual case will be very close to the ideal case.

6.3.3.5 Maximum Duty Ratio

Recalling that the boost and buck-boost types of converters cannot operate with a switching signal u that has a duty ratio $d = 1$, a small protective device is required to ensure that the duty ratio of the controller's output is always $d < 1$.

6.3.3.6 Soft-Starting and Over-Current Protection Devices

The design of the PWM-based SM controller does not take into consideration the issues of soft-starting and over-current protection. In principle, the soft-starting and over-current protection circuits required in this controller is identical to that of the conventional linear voltage-mode controller.

6.4 Simulation Results and Discussions

The derived PWM-based controllers' equations in Table 6.3 are verified through computer simulation.[1] It should be noted that the simulation models used in this chapter and the next chapter do not take into consideration non-idealities such as propagation time delay and slew rates of practical components. Hence, the simulation results may differ slightly from the experimental results obtained in Chapter 8.

6.4.1 Buck Converter

Table 11.1 shows the specifications of the buck converter used in this section.

---

[1]The simulation is performed using Matlab/Simulink. The step size taken for all simulations is 10 ns.

TABLE 6.4

Specifications of buck converter.

Description	Parameter	Nominal Value
Input voltage	v_i	24 V
Capacitance	C	150 μF
Capacitor ESR	C_r	21 mΩ
Inductance	L	22 μH
Inductor resistance	l_r	0.12 Ω
Switching frequency	f_S	200 kHz
Minimum load resistance	$r_{L(min)}$	3 Ω
Maximum load resistance	$r_{L(max)}$	12 Ω
Desired output voltage	V_{od}	12 V

TABLE 6.5

Specifications of boost converter.

Description	Parameter	Nominal Value
Input voltage	v_i	12 V
Capacitance	C	1000 μF
Capacitor ESR	c_r	36 mΩ
Inductance	L	50 μH
Inductor resistance	l_r	0.14 Ω
Switching frequency	f_S	200 kHz
Minimum load resistance	$r_{L(min)}$	48 Ω
Maximum load resistance	$r_{L(max)}$	192 Ω
Desired output voltage	V_{od}	48 V

6.4.1.1 Steady-State Performance

Figure 6.3 shows the steady-state behavior of the PWM-based SMVC buck converter operating at nominal condition $v_i = 24$ V and $r_L = 3$ Ω. The peak-to-peak inductor ripple current is about 1.6 A and the output voltage ripple is about 30 mV.

Figures 6.4(a) and 6.4(b) show, respectively, the DC output voltage versus the input voltage and the operating load resistance. For the line variation test, it can be observed that v_o decreases with an increasing v_i. Specifically, the output voltage deviation is -0.502 V (i.e., -0.42% of $v_{o(\text{nominal condition})}$) for the entire input range 18 V $\leq v_i \leq$ 30 V, i.e., the line regulation $\frac{dv_o}{dv_i}$ averages at -4.208 mV/V. For the load variation test, it can be concluded that voltage regulation of the converter is robust to load changes, with only a 6.6 mV deviation (i.e., 0.06% of $v_{o(\text{nominal condition})}$) in v_o for the entire load range 3 $\Omega \leq r_L \leq$ 12 Ω, i.e., the load regulation $\frac{dv_o}{dr_L}$ averages at 0.73 mV/Ω.

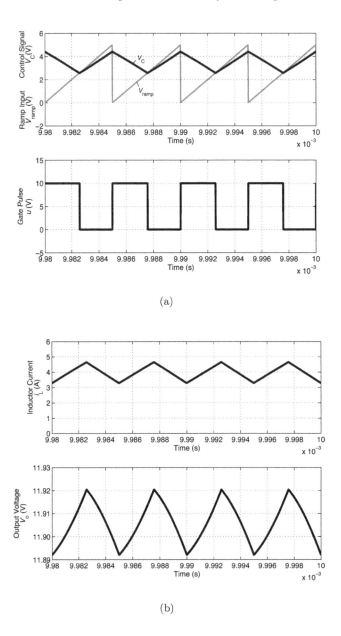

FIGURE 6.3
Waveforms of (a) control signal v_{c}, input ramp v_{ramp}, generated gate pulse u, and (b) inductor current i_{L} and output voltage ripple v_{o} of the PWM-based SMVC buck converter operating at nominal condition $v_{\mathrm{i}} = 24$ V and $r_{\mathrm{L}} = 3 \, \Omega$.

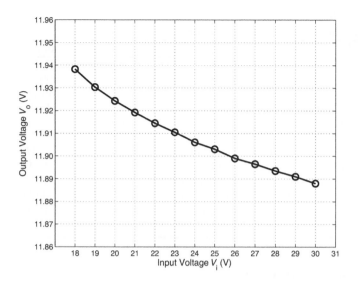

(a) Line variation

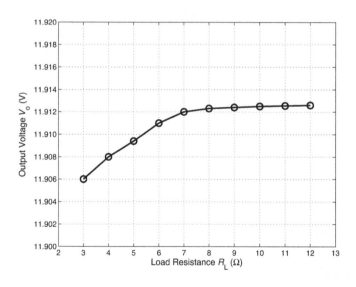

(b) Load variation

FIGURE 6.4
Plot of DC output voltage $\overline{V}_o$ against (a) input voltage v_i and (b) load resistance r_L of the PWM-based SMVC buck converter.

TABLE 6.6

Specifications of buck-boost converter.

Description	Parameter	Nominal Value
Input voltage	v_i	24 V
Capacitance	C	1000 μF
Capacitor ESR	c_r	36 mΩ
Inductance	L	25 μH
Inductor resistance	l_r	0.12 Ω
Switching frequency	f_S	200 kHz
Minimum load resistance (buck)	$r_{L(min)}$	6 Ω
Minimum load resistance (boost)	$r_{L(min)}$	18 Ω
Maximum load resistance (buck)	$r_{L(max)}$	12 Ω
Maximum load resistance (boost)	$r_{L(max)}$	60 Ω
Desired output voltage (buck)	V_{od}	12 V
Desired output voltage (boost)	V_{od}	36 V

6.4.2 Boost Converter

Table 9.1 shows the specifications of the boost converter used in this section.

6.4.2.1 Steady-State Performance

Figure 6.5 shows the steady-state behavior of the PWM-based SMVC boost converter operating at nominal condition $v_i = 12$ V and $r_L = 48$ Ω. The peak-to-peak inductor ripple current is about 1.0 A and the output voltage ripple is about 175 mV.

Figures 6.6(a) and 6.6(b) show, respectively, the DC output voltage versus the input voltage and the operating load resistance. For the line variation test, it can be observed that v_o increases with an increasing v_i. Specifically, the output voltage deviation is 0.514 V (i.e., 1.07% of $v_{o(nominal\ condition)}$) for the entire input range 8 V $\leq v_i \leq$ 16 V , i.e., the line regulation $\frac{dv_o}{dv_i}$ averages at 64.25 mV/V. For the load variation test, it can be concluded that the voltage regulation of the converter is robust to load changes, with only a -0.133 V deviation (i.e., -0.28% of $v_{o(nominal\ condition)}$) in v_o for the entire load range 48 $\Omega \leq r_L \leq$ 192 Ω, i.e., the load regulation $\frac{dv_o}{dr_L}$ averages at -0.92 mV/Ω.

6.4.3 Buck-Boost Converter

Table 6.6 shows the specifications of the buck-boost converter used in this section. The converter is to be operated in both the buck and boost modes.

6.4.3.1 Steady-State Performance

Figures 6.7 and 6.8 show, respectively, the steady-state behavior of the PWM-based SMVC buck-boost converter operating as step-down and step-up con-

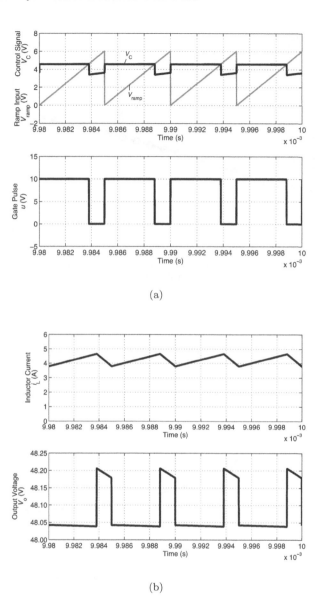

FIGURE 6.5
Waveforms of (a) control signal v_c, input ramp v_{ramp}, generated gate pulse u, and (b) inductor current i_L and output voltage ripple v_o of the PWM-based SMVC boost converter operating at nominal condition $v_i = 12$ V and $r_L = 48$ Ω.

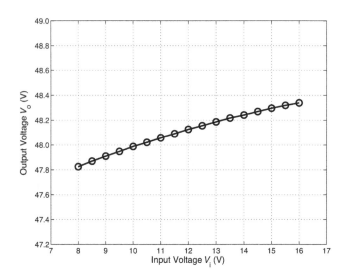

(a) Line variation

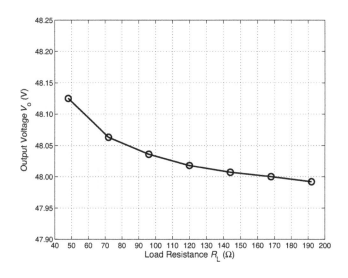

(b) Load variation

FIGURE 6.6
Plot of DC output voltage $\overline{V}_\mathrm{o}$ against (a) input voltage v_i and (b) load resistance r_L of the PWM-based SMVC boost converter.

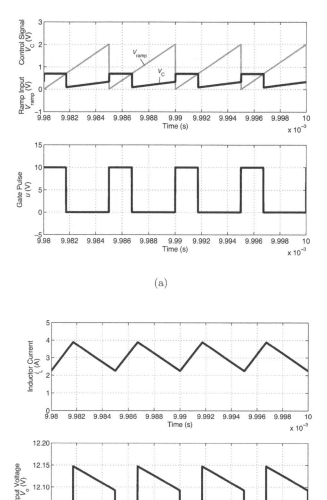

(a)

(b)

FIGURE 6.7

Waveforms of (a) control signal v_c, input ramp v_{ramp}, generated gate pulse u, and (b) inductor current i_L and output voltage ripple v_o of the PWM-based SMVC buck-boost converter operating as step-down converter at nominal condition $v_i = 24$ V, $v_o = 12$ V, and $r_L = 6$ Ω.

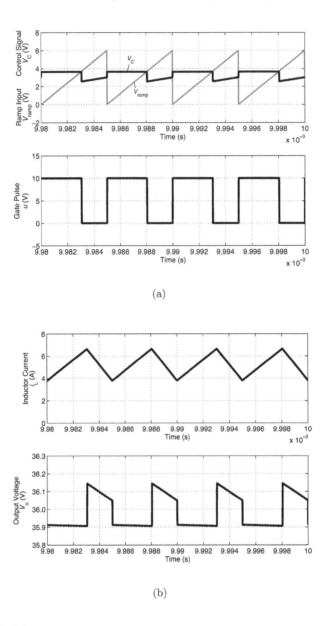

(a)

(b)

FIGURE 6.8
Waveforms of (a) control signal v_c, input ramp v_{ramp}, generated gate pulse u, and (b) inductor current i_L and output voltage ripple v_o (right) of the PWM-based SMVC buck-boost converter operating as step-up converter at nominal condition $v_i = 24$ V, $v_o = 36$ V, and $r_L = 18$ Ω.

verters at their nominal conditions. The corresponding peak-to-peak inductor ripple currents are about 1.5 A and 3.0 A, and the output voltage ripples are about 130 mV and 240 mV.

Figures 6.9(a) and 6.9(b) and Figs. 6.10(a) and 6.10(b) show, respectively, the DC output voltage versus the input voltage and the operating load resistance for both the step-down and step-up conversions. For the step-down conversion (i.e., $v_o = 12$ V), v_o increases with an increasing v_i. The output voltage deviation is 0.152 V (i.e., 1.26% of $v_{o(\text{nominal condition})}$) for the entire input range 18 V $\leq v_i \leq 30$ V, i.e., the line regulation $\frac{dv_o}{dv_i}$ averages at 12.67 mV/V. Also, the voltage variation of the converter to the load change is -0.114 V (i.e., -0.94% of $v_{o(\text{nominal condition})}$) in v_o for the entire load range 6 $\Omega \leq r_L \leq 18$ Ω, i.e., the load regulation $\frac{dv_o}{dr_L}$ averages at -12.67 mV/Ω. For the step-up conversion (i.e., $v_o = 36$ V), v_o also increases with an increasing v_i. The output voltage deviation is 0.443 V (i.e., 1.23% of $v_{o(\text{nominal condition})}$) for the entire input range 18 V $\leq v_i \leq 30$ V, i.e., the line regulation $\frac{dv_o}{dv_i}$ averages at 36.92 mV/V. Also, the voltage variation of the converter to the load change is -0.130 V (i.e., -0.36% of $v_{o(\text{nominal condition})}$) in v_o for the entire load range 18 $\Omega \leq r_L \leq 60$ Ω, i.e., the load regulation $\frac{dv_o}{dr_L}$ averages at -3.10 mV/Ω.

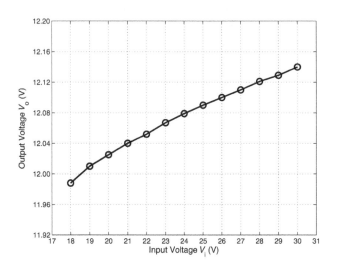

(a) Line variation

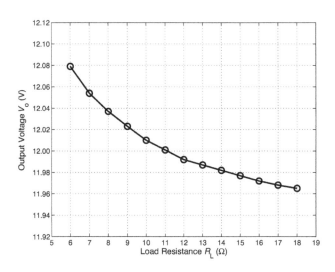

(b) Load variation

FIGURE 6.9
Plot of DC output voltage $\overline{V}_o$ against (a) input voltage v_i and (b) load resistance r_L of the PWM-based SMVC buck-boost converter operating as step-down converter.

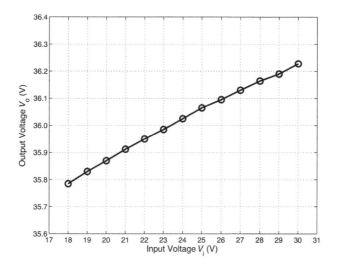

(a) Line variation

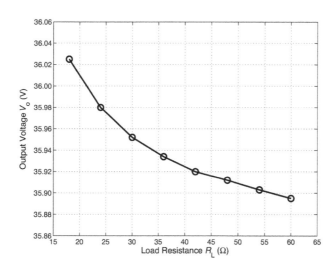

(b) Load variation

FIGURE 6.10
Plot of DC output voltage $\overline{V}_o$ against (a) input voltage v_i and (b) load resistance r_L of the PWM-based SMVC buck-boost converter operating as step-up converter.

7

General Approach for Deriving PWM-Based Sliding Mode Controller for Power Converters in Discontinuous Conduction Mode

CONTENTS

7.1 Introduction

DC–DC converters can operate in either continuous conduction mode (CCM) or discontinuous conduction mode (DCM), depending upon the choice of the switching frequency, and the relative sizes of the load and the inductive storage. The design of pulse-width modulation (PWM)-based SM controllers for DC–DC converters in CCM operation has been thoroughly discussed in the previous chapter, in which the system models, SM control laws, and computer simulations are presented in detail. In practice, DCM operation enjoys a faster transient response at the expense of higher device stresses. It is still a popular operating mode for low power applications and its practical importance should not be overlooked. However, the results presented in the previous chapter are not applicable to DC–DC converters operating in DCM because of the fundamental difference in the dynamical property between the two operations. Thus, if PWM-based SM controllers are to be used for DC–DC converters in DCM operation, system models and control laws have to be redeveloped. The main difference lies in the structural composition of the respective converter models for CCM converters (bilinear) and DCM converters (trilinear).

In this chapter, we continue with our pursuit of a unified approach for SM controller design for DC–DC converters, and in particular we derive the system models and SM control laws for the DCM converter counterparts. In addition to the derivation of system models and control laws, a comparative study is also made to assess the performance of SM-controlled DC–DC converters in DCM operation when the controllers are derived assuming a DCM operation and a CCM operation. Computer simulations are used for evaluation and verification purposes.

It is worth mentioning that the fundamental principle behind the described design approach can be used for the design of multi-switches/multi-structured converter systems, e.g., power factor correction (PFC) and parallel-connected converters in either CCM and DCM operations. However, to avoid obscuring the essentials, our discussions here are limited to the basic DC–DC converters, namely, buck, boost, and buck-boost converters.

7.2 State-Space Converter Model of the DC–DC Converters under DCM

As earlier remarked, the state-space model of a DCM converter is adopted in this design approach. The difference between this model and the model of a CCM converter is the addition of the zero-inductor-current stage. Here, the DCM converter models will be developed by introducing additional terms, known as the virtual switching components, into the models. Such an analogy is only a theoretical representation. There is no additional physical switch needed in the converter circuit.

For the case of buck, boost, and buck-boost converters, the virtual switching components u_L and u_B, on top of the actual physical switching component u where logic 1 and 0 represent the "ON" and "OFF" stages of the actual

TABLE 7.1
State-space model of buck, boost, and buck-boost converters in DCM.

Converter	$\frac{di_{Lr}}{dt}$	$\frac{di_{Lf}}{dt}$	i_L	v_o
Buck	$\frac{v_i - v_o}{L}$	$-\frac{v_o}{L}$	$\frac{1}{L} \int [v_i u - v_o u_L]\, dt$	$\frac{1}{C} \int [i_L - i_r]\, dt$
Boost	$\frac{v_i}{L}$	$\frac{v_i - v_o}{L}$	$\frac{1}{L} \int [v_i u + (v_i - v_o)u_B]\, dt$	$\frac{1}{C} \int [i_L \bar{u} - i_r]\, dt$
Buck-Boost	$\frac{v_i}{L}$	$-\frac{v_o}{L}$	$\frac{1}{L} \int [v_i u - v_o u_B]\, dt$	$\frac{1}{C} \int [i_L \bar{u} - i_r]\, dt$

power switch, are introduced into the models:

$$u_L = \begin{cases} 1 = \text{`ON'} & \text{when } i_L > 0 \\ 0 = \text{`OFF'} & \text{when } i_L = 0 \end{cases}; \tag{7.1}$$

and

$$u_B = \begin{cases} 1 = \text{`ON'} & \text{when } i_L > 0 \text{ and } u = 0 \\ 0 = \text{`OFF'} & \text{when } i_L = 0 \text{ and } u = 0 \end{cases}. \tag{7.2}$$

Here, the condition (7.1) infers that u_L inherits a logic state 1 whenever inductor current i_L is conducting, and condition (7.2) infers that u_B inherits a logic state 1 only when inductor current is conducting and the power switch is off ($u = 0$).

Figure 7.1 illustrates the typical inductor current behavior of a DC–DC converter in both the CCM and DCM operations. The respective rate of change of inductor currents (i.e., $\frac{di_{Lr}}{dt}$ and $\frac{di_{Lf}}{dt}$) and state-space inductor current i_L descriptions for the buck, boost, and buck-boost converters are stated in Table 7.1. It can be seen that in CCM operation, u_B is always 1 whenever $u = 0$, i.e., $u_B = \bar{u}$. Thus, a model derived for the DCM operation, which involves u_B, can be easily transformed to a model for the CCM operation, through the substitution of the condition $u_B = \bar{u}$.

The relationship between the converters' state-space output voltage v_o, inductor current i_L, switching signal $u/\bar{u}$, and load current i_r are also illustrated in Table 7.1.

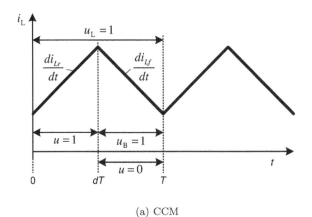

(a) CCM

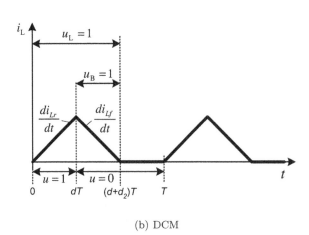

(b) DCM

FIGURE 7.1
Typical inductor current behavior of a DC–DC converter in (a) CCM and (b) DCM operations.

7.3   The Approach

In this section, the modeling method and the detailed procedures for designing the PWM-based SM controllers for DC–DC converters in DCM operation is provided.

7.3.1 System Modeling

The same set of proportional-integral-derivative (PID) SMVC DC–DC converters discussed in the previous chapter is adopted here. Figure 6.1 shows the schematic diagram of the converters. Similar to CCM methodology, we first express the control variables x of the PID SMVC converters in the form:

$$x = \begin{bmatrix} x_1 \\ x_2 \\ x_3 \end{bmatrix} = \begin{bmatrix} V_{ref} - \beta v_o \\ \dfrac{d(V_{ref} - \beta v_o)}{dt} \\ \int (V_{ref} - \beta v_o) dt \end{bmatrix} \tag{7.3}$$

where x_1, x_2, and x_3 represent the *voltage error*, the *voltage error dynamics* (or the rate of change of voltage error), and the *integral of voltage error*, respectively.

Substitution of the converters' behavioral models under DCM (in Table 7.1) into (7.3) produces the following control variable descriptions: x_{buck}, x_{boost}, and $x_{buck-boost}$ for buck, boost, and buck-boost converter, respectively.

$$x_{buck} = \begin{bmatrix} x_1 &=& V_{ref} - \beta v_o \\ x_2 &=& \dfrac{\beta v_o}{r_L C} + \displaystyle\int \dfrac{\beta(v_o u_L - v_i u)}{LC} dt \\ x_3 &=& \int x_1 dt \end{bmatrix} ; \tag{7.4}$$

$$x_{boost} = \begin{bmatrix} x_1 &=& V_{ref} - \beta v_o \\ x_2 &=& \dfrac{\beta v_o}{r_L C} + \displaystyle\int \dfrac{\beta(v_o u_B - v_i u_B)\bar{u}}{LC} dt \\ x_3 &=& \int x_1 dt \end{bmatrix} ; \tag{7.5}$$

$$x_{buck-boost} = \begin{bmatrix} x_1 &=& V_{ref} - \beta v_o \\ x_2 &=& \dfrac{\beta v_o}{r_L C} + \displaystyle\int \dfrac{\beta v_o u_B \bar{u}}{LC} dt \\ x_3 &=& \int x_1 dt \end{bmatrix} . \tag{7.6}$$

Next, the time differentiation of equations (7.4), (7.5), and (7.6) produces the state-space descriptions required for the controller design of the respective converter.

For buck converter:

$$\begin{bmatrix} \dot{x}_1 \\ \dot{x}_2 \\ \dot{x}_3 \end{bmatrix} = \begin{bmatrix} 0 & 1 & 0 \\ 0 & -\frac{1}{r_L C} & 0 \\ 1 & 0 & 0 \end{bmatrix} \begin{bmatrix} x_1 \\ x_2 \\ x_3 \end{bmatrix} + \begin{bmatrix} 0 \\ -\frac{\beta v_i}{LC} \\ 0 \end{bmatrix} u + \begin{bmatrix} 0 \\ \frac{\beta v_o}{LC} u_L \\ 0 \end{bmatrix} , \tag{7.7}$$

For boost converter:

$$\begin{bmatrix} \dot{x}_1 \\ \dot{x}_2 \\ \dot{x}_3 \end{bmatrix} = \begin{bmatrix} 0 & 1 & 0 \\ 0 & -\frac{1}{r_L C} & 0 \\ 1 & 0 & 0 \end{bmatrix} \begin{bmatrix} x_1 \\ x_2 \\ x_3 \end{bmatrix} + \begin{bmatrix} 0 \\ \frac{\beta v_o}{LC} u_B - \frac{\beta v_i}{LC} u_B \\ 0 \end{bmatrix} \bar{u} . \tag{7.8}$$

TABLE 7.2
Descriptions of SMVC buck, boost, and buck-boost converters operating in DCM.

Type of Converter	**A**	**B**	**D**	v
Buck	$\begin{bmatrix} 0 & 1 & 0 \\ 0 & -\frac{1}{r_{\mathrm{L}}C} & 0 \\ 1 & 0 & 0 \end{bmatrix}$	$\begin{bmatrix} 0 \\ -\frac{\beta v_{\mathrm{i}}}{LC} \\ 0 \end{bmatrix}$	$\begin{bmatrix} 0 \\ \frac{\beta v_o}{LC}u_{\mathrm{L}} \\ 0 \end{bmatrix}$	u
Boost	$\begin{bmatrix} 0 & 1 & 0 \\ 0 & -\frac{1}{r_{\mathrm{L}}C} & 0 \\ 1 & 0 & 0 \end{bmatrix}$	$\begin{bmatrix} 0 \\ \frac{\beta v_o}{LC}u_{\mathrm{B}} - \frac{\beta v_{\mathrm{i}}}{LC}u_{\mathrm{B}} \\ 0 \end{bmatrix}$	$\begin{bmatrix} 0 \\ 0 \\ 0 \end{bmatrix}$	$\bar{u}$
Buck-Boost	$\begin{bmatrix} 0 & 1 & 0 \\ 0 & -\frac{1}{r_{\mathrm{L}}C} & 0 \\ 1 & 0 & 0 \end{bmatrix}$	$\begin{bmatrix} 0 \\ \frac{\beta v_o}{LC}u_{\mathrm{B}} \\ 0 \end{bmatrix}$	$\begin{bmatrix} 0 \\ 0 \\ 0 \end{bmatrix}$	$\bar{u}$

For buck-boost converter:

$$\begin{bmatrix} \dot{x}_1 \\ \dot{x}_2 \\ \dot{x}_3 \end{bmatrix} = \begin{bmatrix} 0 & 1 & 0 \\ 0 & -\frac{1}{r_{\mathrm{L}}C} & 0 \\ 1 & 0 & 0 \end{bmatrix} \begin{bmatrix} x_1 \\ x_2 \\ x_3 \end{bmatrix} + \begin{bmatrix} 0 \\ \frac{\beta v_o}{LC}u_{\mathrm{B}} \\ 0 \end{bmatrix} \bar{u}. \tag{7.9}$$

Here, $\bar{u} = 1 - u$ is the inverse logic of u, which is used for modeling the boost and buck-boost topologies. Rearrangement of the state-space descriptions (7.7), (7.8), and (7.9) into the standard form gives

$$\dot{x} = \mathbf{A}x + \mathbf{B}v + \mathbf{D} \tag{7.10}$$

where $v = u$ or $\bar{u}$ (depending on topology). Results are summarized in the tabulated format shown in Table 7.2.

Inspection of the equations verifies the presence of the three state-space structures (trilinear structure) in each converter. In the case of the buck converter, one structure exists when $u = 1$ and $u_{\mathrm{L}} = 1$, another exists when $u = 0$ and $u_{\mathrm{L}} = 1$, and a third exists when $u = 0$ and $u_{\mathrm{L}} = 0$. For the boost and buck-boost converters, one structure exists when $u = 1$ and $u_{\mathrm{B}} = 0$, another exists when $u = 0$ and $u_{\mathrm{B}} = 1$, and a third exists when $u = 0$ and $u_{\mathrm{B}} = 0$.

7.3.2 Controller Design

With the system modeling completed, the next stage is the controller design. We adopt the same PID SM voltage controller as in the previous chapter, which has a control law that employs a switching function

$$
u = \begin{cases} 1 & \text{when } S > 0 \\ 0 & \text{when } S < 0 \end{cases} \tag{7.11}
$$

where S is the instantaneous state trajectory, and is described as

$$
S = \alpha_1 x_1 + \alpha_2 x_2 + \alpha_3 x_3 = \boldsymbol{J}^{\mathrm{T}} \boldsymbol{x}, \tag{7.12}
$$

with $\boldsymbol{J}^{\mathrm{T}} = [\alpha_1 \ \alpha_2 \ \alpha_3]$ and α_1, α_2, and α_3 representing the sliding coefficients.

7.3.2.1 Derivation of Existence Conditions

Although the exact controller is used, the method of obtaining the existence conditions is slightly more complicated in this case than in the case of the CCM operation. This is because in the DCM operation, the trilinear structure behavior of the system results in a trilinear structure state trajectory S. The difference between CCM and DCM operations is illustrated in Fig. 7.2. Typically, for CCM operation, the trajectory in each switching sub-interval must comply with the existence condition so that it would stay within a small vicinity of the sliding plane. For DCM operation, which has a trilinear structure trajectory, the trajectory in two of the three sub-intervals should comply with the existence condition in order to ensure that it would stay within a small vicinity of the sliding plane. However, if the trajectory in all sub-intervals comply with the existence conditions, the system will be more stable.

Our next step is to derive the existence conditions of SM control operation for the converters in DCM operation. Similar to converters in CCM operation, the basic principle is to ensure that the trajectory obeys the local reachability condition in each subinterval, i.e.,

$$
\lim_{S \to 0} S \cdot \dot{S} < 0. \tag{7.13}
$$

This can be expressed as

$$
\begin{cases} \dot{S}_{S \to 0^+} = \boldsymbol{J}^{\mathrm{T}} \boldsymbol{A} \boldsymbol{x} + \boldsymbol{J}^{\mathrm{T}} \boldsymbol{B} v_{S \to 0^+} + \boldsymbol{J}^{\mathrm{T}} \boldsymbol{D} < 0 \\ \dot{S}_{S \to 0^-} = \boldsymbol{J}^{\mathrm{T}} \boldsymbol{A} \boldsymbol{x} + \boldsymbol{J}^{\mathrm{T}} \boldsymbol{B} v_{S \to 0^-} + \boldsymbol{J}^{\mathrm{T}} \boldsymbol{D} > 0 \end{cases} . \tag{7.14}
$$

The specific conditions for the existence of SM control operation for the respective converters are derived as follows:

i) Buck converter

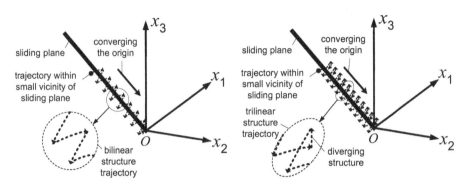

(a) Trajectory in CCM operation   (b) Trajectory in DCM operation

FIGURE 7.2
Graphical representations of state trajectory's behavior in SM control process:
(a) Trajectory behavior in CCM operation—illustrating how bilinear structure
trajectory are maintained within a small vicinity from the sliding plane; and
(b) Trajectory behavior in CCM operation—illustrating how trilinear structure trajectory are maintained within a small vicinity from the sliding plane,
even though one of its structure is not directing the trajectory toward the
sliding plane.

  − Case 1: $S \to 0^+$, $\dot{S} < 0$:
    Substitution of $v_{S \to 0^+} = u = 1$, $u_L = 0$, and the matrices in Table
    7.2 into (7.14) gives

$$-\beta L \left(\frac{\alpha_1}{\alpha_2} - \frac{1}{r_L C} \right) i_C + LC \frac{\alpha_3}{\alpha_2} (V_{\text{ref}} - \beta v_o) + \beta v_o - \beta v_i < 0. \quad (7.15)$$

  − Case 2: $S \to 0^-$, $\dot{S} > 0$:
    Substitution of $v_{S \to 0^-} = u = 0$, $u_L = 1$, and the matrices in Table
    7.2 into (7.14) gives

$$-\beta L \left(\frac{\alpha_1}{\alpha_2} - \frac{1}{r_L C} \right) i_C + LC \frac{\alpha_3}{\alpha_2} (V_{\text{ref}} - \beta v_o) + \beta v_o > 0. \quad (7.16)$$

    Substitution of $v_{S \to 0^-} = u = 0$, $u_L = 0$, and the matrices in Table
    7.2 into (7.14) gives

$$-\beta L \left(\frac{\alpha_1}{\alpha_2} - \frac{1}{r_L C} \right) i_C + LC \frac{\alpha_3}{\alpha_2} (V_{\text{ref}} - \beta v_o) > 0. \quad (7.17)$$

It can be observed that (7.17) is a stricter constraint than (7.16).

Hence, the combination of (7.15) and (7.17) gives the simplified existence condition

$$0 < -\beta L \left(\frac{\alpha_1}{\alpha_2} - \frac{1}{r_L C} \right) i_C + LC \frac{\alpha_3}{\alpha_2} (V_{\text{ref}} - \beta v_o) < \beta (v_i - v_o). \quad (7.18)$$

ii) Boost converter

- Case 1: $S \to 0^+$, $\dot{S} < 0$:
  Substitution of $v_{S \to 0^+} = \bar{u} = 0$, $u_B = 0$, and the matrices in Table 7.2 into (7.14) gives

$$-\beta L \left(\frac{\alpha_1}{\alpha_2} - \frac{1}{r_L C} \right) i_C + LC \frac{\alpha_3}{\alpha_2} (V_{\text{ref}} - \beta v_o) < 0. \quad (7.19)$$

- Case 2: $S \to 0^-$, $\dot{S} > 0$:
  Substitution of $v_{S \to 0^-} = \bar{u} = 1$, $u_B = 1$, and the matrices in Table 7.2 into (7.14) gives

$$-\beta L \left(\frac{\alpha_1}{\alpha_2} - \frac{1}{r_L C} \right) i_C + LC \frac{\alpha_3}{\alpha_2} (V_{\text{ref}} - \beta v_o) + \beta v_o - \beta v_i > 0. \quad (7.20)$$

  Substitution of $v_{S \to 0^-} = \bar{u} = 1$, $u_B = 0$, and the matrices in Table 7.2 into (7.14) gives

$$-\beta L \left(\frac{\alpha_1}{\alpha_2} - \frac{1}{r_L C} \right) i_C + LC \frac{\alpha_3}{\alpha_2} (V_{\text{ref}} - \beta v_o) > 0. \quad (7.21)$$

For this converter, it can be seen that (7.21) contradicts (7.19). Since only two existence conditions must be complied (one for $S \to 0^+$ and another for $S \to 0^-$) for the system trajectory to be directed toward the sliding surface (i.e., into SM operation), one of the conditions in Case 2 can be neglected. Hence, only (7.19) and (7.20) are considered. The combination of these conditions gives

$$0 < \beta L \left(\frac{\alpha_1}{\alpha_2} - \frac{1}{r_L C} \right) i_C - LC \frac{\alpha_3}{\alpha_2} (V_{\text{ref}} - \beta v_o) < \beta (v_o - v_i). \quad (7.22)$$

Failure in satisfying (7.21) may cause the state trajectory of the system to move away from the sliding surface during the zero-inductor-current stage. However, as explained, the abidance of (7.22) will still ensure the occurrence of SM operation.

iii) Buck-boost converter

- Case 1: $S \to 0^+$, $\dot{S} < 0$:
  Substitution of $v_{S \to 0^+} = \bar{u} = 0$, $u_B = 0$, and the matrices in Table 7.2 into (7.14) gives

$$-\beta L \left(\frac{\alpha_1}{\alpha_2} - \frac{1}{r_L C} \right) i_C + LC \frac{\alpha_3}{\alpha_2} (V_{\text{ref}} - \beta v_o) < 0. \quad (7.23)$$

TABLE 7.3

Existence conditions of buck, boost, and buck-boost converters operating in DCM.

Buck	$0 < -\beta L \left(\frac{\alpha_1}{\alpha_2} - \frac{1}{r_\mathrm{L} C} \right) i_C + LC \frac{\alpha_3}{\alpha_2} (V_\mathrm{ref} - \beta v_\mathrm{o}) < \beta v_\mathrm{i} - \beta v_\mathrm{o}$
Boost	$0 < \beta L \left(\frac{\alpha_1}{\alpha_2} - \frac{1}{r_\mathrm{L} C} \right) i_C - LC \frac{\alpha_3}{\alpha_2} (V_\mathrm{ref} - \beta v_\mathrm{o}) < \beta(v_\mathrm{o} - v_\mathrm{i})$ (same as CCM)
Buck-Boost	$0 < \beta L \left(\frac{\alpha_1}{\alpha_2} - \frac{1}{r_\mathrm{L} C} \right) i_C - LC \frac{\alpha_3}{\alpha_2} (V_\mathrm{ref} - \beta v_\mathrm{o}) < \beta v_\mathrm{o}$ (same as CCM)

- Case 2: $S \to 0^-$, $\dot{S} > 0$:

    Substitution of $v_{S \to 0^-} = \bar{u} = 1$, $u_\mathrm{B} = 1$, and the matrices in Table 7.2 into (7.14) gives

$$-\beta L \left(\frac{\alpha_1}{\alpha_2} - \frac{1}{r_\mathrm{L} C} \right) i_C + LC \frac{\alpha_3}{\alpha_2} (V_\mathrm{ref} - \beta v_\mathrm{o}) + \beta v_\mathrm{o} > 0. \quad (7.24)$$

    Substitution of $v_{S \to 0^-} = \bar{u} = 1$, $u_\mathrm{B} = 0$, and the matrices in Table 7.2 into (7.14) gives

$$-\beta L \left(\frac{\alpha_1}{\alpha_2} - \frac{1}{r_\mathrm{L} C} \right) i_C + LC \frac{\alpha_3}{\alpha_2} (V_\mathrm{ref} - \beta v_\mathrm{o}) > 0. \quad (7.25)$$

For this converter, (7.25) contradicts (7.23). Hence, only (7.23) and (7.24) are considered. The combination of these conditions gives

$$0 < \beta L \left(\frac{\alpha_1}{\alpha_2} - \frac{1}{r_\mathrm{L} C} \right) i_C - LC \frac{\alpha_3}{\alpha_2} (V_\mathrm{ref} - \beta v_\mathrm{o}) < \beta v_\mathrm{o}. \quad (7.26)$$

The derived existence conditions for all three converters are tabulated in Table 7.3. The selection of sliding coefficients for the controller of each converter must comply with the stated inequalities.

7.3.2.2 Derivation of Control Equations for PWM-Based Controller

The derivation of the PWM-based controllers is similar to that for the CCM converters described in the previous chapter. First, the SM control law is translated to its equivalent control function. Then, it is translated to the duty-ratio function of the pulse-width modulator.

i) Buck converter

Equating $\dot{S} = \boldsymbol{J}^{\mathrm{T}}\mathbf{A}\boldsymbol{x} + \boldsymbol{J}^{\mathrm{T}}\mathbf{B}u_{\mathrm{eq}} + \mathbf{D} = 0$ yields the equivalent control function

$$u_{\mathrm{eq}} = -\left[\boldsymbol{J}^{\mathrm{T}}\mathbf{B}\right]^{-1}\boldsymbol{J}^{\mathrm{T}}[\mathbf{A}\boldsymbol{x} + \mathbf{D}] \tag{7.27}$$

$$= -\frac{\beta L}{\beta v_{\mathrm{i}}}\left(\frac{\alpha_1}{\alpha_2} - \frac{1}{r_{\mathrm{L}}C}\right)i_C + \frac{\alpha_3 LC}{\alpha_2 \beta v_{\mathrm{i}}}\left(V_{\mathrm{ref}} - \beta v_{\mathrm{o}}\right) + \frac{v_{\mathrm{o}}}{v_{\mathrm{i}}}u_{\mathrm{Leq}}$$

where both u_{eq} and u_{Leq} are continuous and bounded by 0 and 1. Specifically, the equivalent control function u_{eq} is a smooth function of the discrete input function u, and u_{Leq} is the resulting equivalent component of the discrete virtual switching component u_{L}.

The equivalent control function is first derived by substituting (7.27) into $0 < u_{\mathrm{eq}} < 1$, which gives

$$0 < u_{\mathrm{eq}} = -\frac{\beta L}{\beta v_{\mathrm{i}}}\left(\frac{\alpha_1}{\alpha_2} - \frac{1}{r_{\mathrm{L}}C}\right)i_C \tag{7.28}$$

$$+ \frac{\alpha_3 LC}{\alpha_2 \beta v_{\mathrm{i}}}\left(V_{\mathrm{ref}} - \beta v_{\mathrm{o}}\right) + \frac{v_{\mathrm{o}}}{v_{\mathrm{i}}}u_{\mathrm{Leq}} < 1.$$

Multiplication of the inequality by βv_{i} gives

$$0 < u_{\mathrm{eq}}{}^* = -\beta L\left(\frac{\alpha_1}{\alpha_2} - \frac{1}{r_{\mathrm{L}}C}\right)i_C \tag{7.29}$$

$$+ LC\frac{\alpha_3}{\alpha_2}\left(V_{\mathrm{ref}} - \beta v_{\mathrm{o}}\right) + \beta v_{\mathrm{o}}u_{\mathrm{Leq}} < \beta v_{\mathrm{i}}$$

Next, the function u_{Leq} can be derived using the usual average model of the buck converter. First, for the buck converter in DCM, the duty ratio is given by

$$d_{\mathrm{DCM}} = \sqrt{\frac{v_{\mathrm{o}}{}^2}{v_{\mathrm{i}}(v_{\mathrm{i}} - v_{\mathrm{o}})}\frac{2L}{r_{\mathrm{L}}T}}. \tag{7.30}$$

Also, the duty ratio for the turn-off period in DCM operation is

$$d_2 = \frac{v_{\mathrm{i}} - v_{\mathrm{o}}}{v_{\mathrm{o}}}d_{\mathrm{DCM}} = \sqrt{\frac{v_{\mathrm{i}} - v_{\mathrm{o}}}{v_{\mathrm{i}}}\frac{2L}{r_{\mathrm{L}}T}}. \tag{7.31}$$

Since the maximum time duration that $u_L = 1$ can exist is T (during CCM operation), and in DCM operation $u_L = 1$ is applied for the time duration $(d_{\mathrm{DCM}} + d_2)T$, the equivalent virtual switching component applied to the system in DCM operation can be described as

$$u_{\mathrm{Leq}} = \frac{(d_{\mathrm{DCM}} + d_2)T}{T} = \sqrt{\frac{v_{\mathrm{i}}}{v_{\mathrm{i}} - v_{\mathrm{o}}}\frac{2L}{r_{\mathrm{L}}T}} \tag{7.32}$$

Finally, the translation of the equivalent control function (7.29) to the duty ratio d, where $0 < d = \frac{v_c}{\hat{v}_{ramp}} < 1$, gives the following relationships for the control signal v_c and ramp signal $\hat{v}_{ramp}$:

$$v_c = u_{eq}{}^* = -\beta L \left(\frac{\alpha_1}{\alpha_2} - \frac{1}{r_L C} \right) i_C \tag{7.33}$$
$$+ \frac{\alpha_3}{\alpha_2} LC \left(V_{ref} - \beta v_o \right) + \beta v_o u_{Leq}$$

and

$$\hat{v}_{ramp} = \beta v_i \tag{7.34}$$

for the practical implementation of the PWM-based SM controller.

ii) Boost converter

Equating $\dot{S} = \boldsymbol{J}^T \boldsymbol{A} \boldsymbol{x} + \boldsymbol{J}^T \boldsymbol{B} \bar{u}_{eq} = 0$ yields the equivalent control function

$$\bar{u}_{eq} = -\left[\boldsymbol{J}^T \boldsymbol{B} \right]^{-1} \boldsymbol{J}^T \boldsymbol{A} \boldsymbol{x} \tag{7.35}$$
$$= \frac{\beta L}{\beta \left(v_o - v_i \right) u_{Beq}} \left(\frac{\alpha_1}{\alpha_2} - \frac{1}{r_L C} \right) i_C - \frac{\alpha_3 LC}{\alpha_2 \beta \left(v_o - v_i \right) u_{Beq}} \left(V_{ref} - \beta v_o \right)$$

where both $\bar{u}_{eq}$ and u_{Beq} are continuous and bounded by 0 and 1. Here, the equivalent control function $\bar{u}_{eq}$ is a smooth function of the discrete input function $\bar{u}$, and u_{Beq} is the resulted equivalent component of the discrete virtual switching component u_B.

The equivalent control function is first derived by substituting (7.35) into $0 < \bar{u}_{eq} < 1$, which gives

$$0 < \bar{u}_{eq} = \frac{\beta L}{\beta \left(v_o - v_i \right) u_{Beq}} \left(\frac{\alpha_1}{\alpha_2} - \frac{1}{r_L C} \right) i_C$$
$$- \frac{\alpha_3 LC}{\alpha_2 \beta \left(v_o - v_i \right) u_{Beq}} \left(V_{ref} - \beta v_o \right) < 1. \tag{7.36}$$

Since $u = 1 - \bar{u}$, which also implies $u_{eq} = 1 - \bar{u}_{eq}$, the inequality can be rewritten as

$$0 < u_{eq} = 1 - \frac{\beta L}{\beta \left(v_o - v_i \right) u_{Beq}} \left(\frac{\alpha_1}{\alpha_2} - \frac{1}{r_L C} \right) i_C$$
$$+ \frac{\alpha_3 LC}{\alpha_2 \beta \left(v_o - v_i \right) u_{Beq}} \left(V_{ref} - \beta v_o \right) < 1. \tag{7.37}$$

Multiplication of the inequality by $\beta (v_o - v_i) u_{Beq}$ gives

$$0 < u_{eq}{}^* = -\beta L \left(\frac{\alpha_1}{\alpha_2} - \frac{1}{r_L C} \right) i_C + LC \frac{\alpha_3}{\alpha_2} \left(V_{ref} - \beta v_o \right)$$
$$+ \beta \left(v_o - v_i \right) u_{Beq} < \beta \left(v_o - v_i \right) u_{Beq}. \tag{7.38}$$

Next, the function u_{Beq} can be derived using the usual model of the boost converter. First, for the boost converter in CCM, the duty ratio is given by

$$d_{\text{CCM}} = 1 - \frac{v_{\text{i}}}{v_{\text{o}}}, \tag{7.39}$$

and in DCM, the duty ratio is given by

$$d_{\text{DCM}} = \sqrt{\frac{v_{\text{o}}}{v_{\text{i}}} \left(\frac{v_{\text{o}}}{v_{\text{i}}} - 1 \right) \frac{2L}{r_{\text{L}} T}}. \tag{7.40}$$

In addition, the duty ratio for the turn-off period in DCM operation is given by

$$d_2 = \frac{v_{\text{i}}}{v_{\text{o}} - v_{\text{i}}} d_{\text{DCM}} = \sqrt{\frac{v_{\text{o}}}{v_{\text{o}} - v_{\text{i}}} \frac{2L}{r_{\text{L}} T}}. \tag{7.41}$$

Since the maximum time duration that $u_B = 1$ can exist is $d_{2\max} T = (1 - d_{\text{CCM}}) T$ (during CCM operation), and in DCM operation, $u_B = 1$ is applied for the time duration $d_2 T$, the equivalent virtual switching component applied to the system in DCM operation can be described as

$$u_{\text{Beq}} = \frac{d_2 T}{d_{2\max} T} = \frac{v_{\text{o}}}{v_{\text{i}}} \sqrt{\frac{v_{\text{o}}}{v_{\text{o}} - v_{\text{i}}} \frac{2L}{r_{\text{L}} T}}. \tag{7.42}$$

Finally, the translation of the equivalent control function (7.38) to the duty ratio d, where $0 < d = \frac{v_{\text{c}}}{\hat{v}_{\text{ramp}}} < 1$, gives the following relationships for the control signal v_{c} and ramp signal $\hat{v}_{\text{ramp}}$ where

$$v_{\text{c}} = u_{\text{eq}}{}^* = -\beta L \left(\frac{\alpha_1}{\alpha_2} - \frac{1}{r_{\text{L}} C} \right) i_C + LC \frac{\alpha_3}{\alpha_2} \left(V_{\text{ref}} - \beta v_{\text{o}} \right)$$

$$+ \beta \left(v_{\text{o}} - v_{\text{i}} \right) u_{\text{Beq}} \tag{7.43}$$

and

$$\hat{v}_{\text{ramp}} = \beta \left(v_{\text{o}} - v_{\text{i}} \right) u_{\text{Beq}} \tag{7.44}$$

for the practical implementation of the PWM-based SM controller.

iii) Buck-boost converter

Equating $\dot{S} = \boldsymbol{J}^{\text{T}} \boldsymbol{A} \boldsymbol{x} + \boldsymbol{J}^{\text{T}} \boldsymbol{B} \bar{u}_{\text{eq}} = 0$ yields the equivalent control function

$$\begin{aligned} \bar{u}_{\text{eq}} &= -\left[\boldsymbol{J}^{\text{T}} \boldsymbol{B} \right]^{-1} \boldsymbol{J}^{\text{T}} \boldsymbol{A} \boldsymbol{x} \tag{7.45} \\ &= \frac{\beta L}{\beta v_{\text{o}} u_{\text{Beq}}} \left(\frac{\alpha_1}{\alpha_2} - \frac{1}{r_{\text{L}} C} \right) i_C - \frac{\alpha_3 LC}{\alpha_2 \beta v_{\text{o}} u_{\text{Beq}}} \left(V_{\text{ref}} - \beta v_{\text{o}} \right) \end{aligned}$$

where both $\bar{u}_{\text{eq}}$ and u_{Beq} are continuous and bounded by 0 and 1.

The equivalent control function is derived by substituting (7.45) into $0 < \bar{u}_{eq} < 1$, which gives

$$0 < \bar{u}_{eq} = \frac{\beta L}{\beta v_o u_{Beq}} \left(\frac{\alpha_1}{\alpha_2} - \frac{1}{r_L C} \right) i_C$$

$$- \frac{\alpha_3 L C}{\alpha_2 \beta v_o u_{Beq}} (V_{ref} - \beta v_o) < 1. \tag{7.46}$$

Since $u = 1 - \bar{u}$, which also implies $u_{eq} = 1 - \bar{u}_{eq}$, the inequality can be rewritten as

$$0 < u_{eq} = 1 - \frac{\beta L}{\beta v_o u_{Beq}} \left(\frac{\alpha_1}{\alpha_2} - \frac{1}{r_L C} \right) i_C$$

$$+ \frac{\alpha_3 L C}{\alpha_2 \beta v_o u_{Beq}} (V_{ref} - \beta v_o) < 1. \tag{7.47}$$

Multiplication of the inequality by $\beta v_o u_{Beq}$ gives

$$0 < u_{eq}{}^* = -\beta L \left(\frac{\alpha_1}{\alpha_2} - \frac{1}{r_L C} \right) i_C$$

$$+ L C \frac{\alpha_3}{\alpha_2} (V_{ref} - \beta v_o) + \beta v_o u_{Beq} < \beta v_o u_{Beq}. \tag{7.48}$$

Next, the function u_{Beq} can be derived using the usual model of the buck-boost converter. First, for the buck-boost converter in CCM, the duty ratio is given by

$$d_{CCM} = \frac{v_o}{v_o + v_i}, \tag{7.49}$$

and in DCM, the duty ratio is given by

$$d_{DCM} = \frac{v_o}{v_i} \sqrt{\frac{2L}{r_L T}}. \tag{7.50}$$

Also, the duty ratio for the turn-off period in DCM operation is given as

$$d_2 = \frac{v_i}{v_o} d_{DCM} = \sqrt{\frac{2L}{r_L T}}. \tag{7.51}$$

Since the maximum time duration that $u_B = 1$ can exist is $d_{2max}T = (1 - d_{CCM})T$ (during CCM operation), and in DCM operation, $u_B = 1$ is applied for the time duration $d_2 T$, the equivalent virtual switching component applied to the system in DCM operation can be described as

$$u_{Beq} = \frac{d_2 T}{d_{2max} T} = \left(1 + \frac{v_o}{v_i} \right) \sqrt{\frac{2L}{r_L T}}. \tag{7.52}$$

Finally, the translation of the equivalent control function (7.48) to the

TABLE 7.4

Control equations of PWM-based SMVC buck, boost, and buck-boost converters operating in DCM. K_{p1} and K_{p2} are calculated using $K_{p1} = \beta L \left(\frac{\alpha_1}{\alpha_2} - \frac{1}{r_L C} \right)$ and $K_{p2} = \frac{\alpha_3}{\alpha_2} LC$.

Converter	v_c	$\hat{v}_{ramp}$	Equivalent VSC
Buck	$-K_{p1}i_C + K_{p2}\left(V_{ref} - \beta v_o\right)$ $+\beta v_o u_{Leq}$	βv_i	$u_{Leq} = \sqrt{\frac{v_i}{v_i - v_o} \frac{2L}{r_L T}}$
Boost	$-K_{p1}i_C + K_{p2}\left(V_{ref} - \beta v_o\right)$ $+\beta\left(v_o - v_i\right)u_{Beq}$	$\beta\left(v_o - v_i\right)u_{Beq}$	$u_{Beq} = \frac{v_o}{v_i}\sqrt{\frac{v_o}{v_o - v_i}\frac{2L}{r_L T}}$
Buck-Boost	$-K_{p1}i_C + K_{p2}\left(V_{ref} - \beta v_o\right)$ $+\beta v_o u_{Beq}$	$\beta v_o u_{Beq}$	$u_{Beq} = \left(1 + \frac{v_o}{v_i}\right)\sqrt{\frac{2L}{r_L T}}$

duty ratio d, where $0 < d = \frac{v_c}{\hat{v}_{ramp}} < 1$, gives the following relationships for the control signal v_c and ramp signal $\hat{v}_{ramp}$ where

$$v_c = u_{eq}{}^* = -\beta L \left(\frac{\alpha_1}{\alpha_2} - \frac{1}{r_L C} \right) i_C$$
$$+ LC\frac{\alpha_3}{\alpha_2}\left(V_{ref} - \beta v_o\right) + \beta v_o u_{Beq} \quad (7.53)$$

and

$$\hat{v}_{ramp} = \beta v_o u_{Beq} \quad (7.54)$$

for the practical implementation of the PWM-based SM controller.

The control equations required for the implementation of the respective PWM-based SMVC converter operating in DCM are tabulated in Table 7.4.

7.4   Simulation Results and Discussions

Computer simulations[1] are performed using the derived PWM-based controllers' equations in Table 7.4. Comparisons are also made between the performances of the PWM-based SMVC converters using both the CCM and

---

[1]The simulation is performed using Matlab/Simulink. The step size taken for all simulations is 10 ns.

TABLE 7.5

Specifications of buck converter.

Description	Parameter	Nominal Value
Input voltage	v_i	24 V
Capacitance	C	150 μF
Capacitor ESR	c_r	21 mΩ
Inductance	L	22 μH
Inductor resistance	l_r	0.12 Ω
Switching frequency	f_S	200 kHz
Minimum load resistance	$r_{L(min)}$	3 Ω
Maximum load resistance	$r_{L(max)}$	120 Ω
Desired output voltage	V_{od}	12 V

DCM design approaches. For ease of comprehension, the controller developed from the CCM design approach is termed CCM PWM-based SM controller, and the controller developed from the DCM design approach is termed DCM PWM-based SM controller. The same control parameters (i.e., sliding coefficients) are employed in both controllers.

7.4.1   Buck Converter

Table 7.5 shows the specifications of the buck converter used in this section.

7.4.1.1   Steady-State Performance

Figure 7.3 shows the steady-state behavior of the buck converter with the DCM PWM-based controller. Figure 7.4 shows a plot of the DC output voltage against different operating load resistances. It can be easily observed that for operation in CCM, i.e., for load resistance $r_L < 17.5$ Ω (critical resistance), the CCM PWM-based controller has a better load regulation with a 0.009 V deviation in v_o (i.e., $\frac{dv_o}{dr_L} = 0.62$ mV/Ω) than the DCM PWM-based controller, which has a -0.327 V deviation in v_o (i.e., $\frac{dv_o}{dr_L} = 22.55$ mV/Ω), for the load range 3 $\Omega \leq r_L \leq 17.5$ Ω. For operation in DCM, i.e., at load resistance $r_L > 17.5$ Ω (critical resistance), the DCM PWM-based controller has a better load regulation with a 0.036 V deviation in v_o (i.e., $\frac{dv_o}{dr_L} = 0.35$ mV/Ω) than the CCM PWM-based controller, which has a 0.179 V deviation in v_o (i.e., $\frac{dv_o}{dr_L} = 1.74$ mV/Ω), for the load range 17.5 $\Omega \leq r_L \leq 120$ Ω. These results are in good agreement with our theoretical analysis. Clearly, for converters operating in DCM, the DCM-based controller would outperform the CCM-based controller. Likewise, for converters operating in CCM, the CCM-based controller would outperform the DCM-based controller.

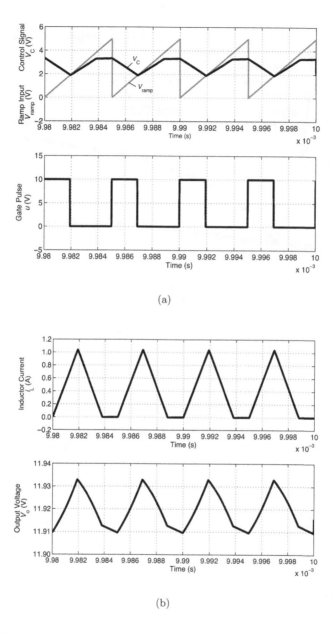

(a)

(b)

FIGURE 7.3
Waveforms of (a) control signal v_c, input ramp v_{ramp}, gate pulse u, and (b) inductor current i_L and output voltage ripple v_o of the PWM-based SMVC buck converter operating in DCM at nominal condition.

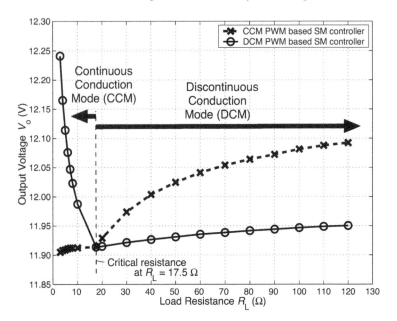

FIGURE 7.4
DC output voltage $\overline{V}_o$ versus load resistance r_L for SMVC buck converter with the CCM PWM-based and the DCM PWM-based controllers.

7.4.1.2 Transient Performance

The dynamic performance of the controllers is studied using a load resistance that alternates between 3 Ω and 12 Ω for CCM operation, and between 30 Ω and 120 Ω for DCM operation. Figures 7.5(a) and 7.5(b) show the output voltage ripple waveforms for both controllers. In CCM operation, the CCM PWM-based controller gives significantly shorter settling times than the DCM PWM-based controller. Conversely, in DCM operation, the DCM PWM-based controller gives significantly shorter settling times than the CCM PWM-based controller. Hence, as with the steady-state operation, each controller outperforms the other in their respective operating modes.

7.4.2 Boost Converter

Table 7.6 shows the specifications of the boost converter used in this section.

7.4.2.1 Steady-State Performance

Figure 7.6 shows the steady-state behavior of the boost converter with the DCM PWM-based controller. It should be noted that the amplitude of the ramp signal will change adaptively with the changes in load resistance. This is

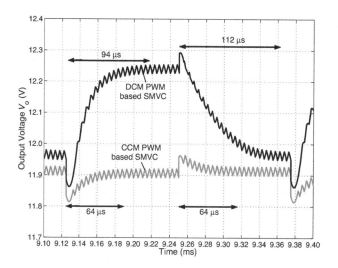

(a) CCM operation; r_L alternating between 3 Ω and 12 Ω

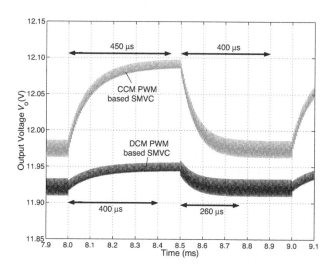

(b) DCM operation; r_L alternating between 30 Ω and 120 Ω

FIGURE 7.5
Waveforms of output voltage v_o of the SMVC buck converter with the CCM PWM-based and the DCM PWM-based controllers operating in (a) CCM and (b) DCM operations.

TABLE 7.6

Specifications of boost converter.

Description	Parameter	Nominal Value
Input voltage	v_i	12 V
Capacitance	C	1000 μF
Capacitor ESR	c_r	36 mΩ
Inductance	L	50 μH
Inductor resistance	l_r	0.14 Ω
Switching frequency	f_S	200 kHz
Minimum load resistance	$r_{L(min)}$	48 Ω
Maximum load resistance	$r_{L(max)}$	800 Ω
Desired output voltage	V_{od}	48 V

due to the equivalent virtual component term in the $\hat{v}_{ramp}$ computation (see (7.44)).

Figure 7.7 shows a plot of the DC output voltage against different operating load resistances. It can be easily observed that for operation in CCM, i.e., at load resistance $r_L < 430\ \Omega$ (critical resistance), the CCM PWM-based controller has a better load regulation with a -0.162 V deviation in v_o (i.e., $\frac{dv_o}{dr_L} = -0.42$ mV/Ω) than the DCM PWM-based controller, which has a -3.310 V deviation in v_o (i.e., $\frac{dv_o}{dr_L} = -8.66$ mV/Ω), for the load range $48\ \Omega \leq r_L \leq 430\ \Omega$. For operation in DCM, i.e., at load resistance $r_L > 430\ \Omega$ (critical resistance), the DCM PWM-based controller gives better load regulation with a 0.376 V deviation in v_o (i.e., $\frac{dv_o}{dr_L} = 1.02$ mV/Ω) than the CCM PWM-based controller, which has a 0.913 V deviation in v_o (i.e., $\frac{dv_o}{dr_L} = 2.47$ mV/Ω), for the load range $430\ \Omega \leq r_L \leq 800\ \Omega$. Similar to the buck converter, the DCM PWM-based controller gives better load regulation property than the CCM PWM-based controller in DCM operation, and vice versa, in the case of controlling the boost converter.

7.4.2.2   Transient Performance

The dynamic performance of the controllers is studied using a load resistance that alternates between 48 Ω and 192 Ω for CCM operation, and between 480 Ω and 800 Ω for DCM operation. Figures 7.8(a) and 7.8(b) show the output voltage ripple waveforms for both controllers. In CCM operation, the CCM PWM-based controller gives significantly shorter settling times than that of the DCM PWM-based controller. Conversely, in DCM operation, the DCM PWM-based controller gives significantly shorter settling times than that of the CCM PWM-based controller.

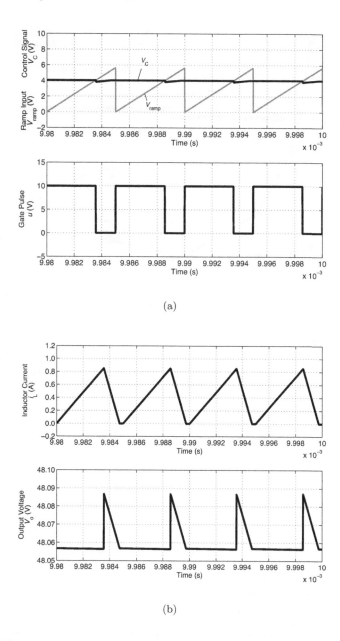

FIGURE 7.6
Waveforms of (a) control signal v_c, input ramp v_{ramp}, gate pulse u, and (b) inductor current i_L and output voltage ripple v_o of the PWM-based SMVC boost converter operating in DCM at nominal condition.

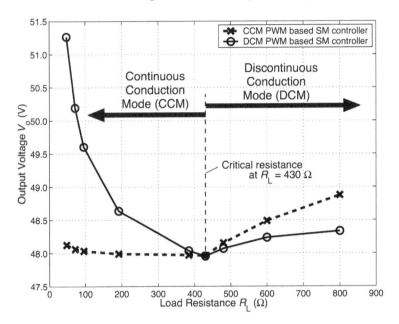

FIGURE 7.7
DC output voltage $\overline{V}_o$ versus load resistance r_L for SMVC boost converter
with the CCM PWM-based and the DCM PWM-based controllers.

TABLE 7.7
Specifications of buck-boost converter.

Description	Parameter	Nominal Value
Input voltage	v_i	24 V
Capacitance	C	1000 μF
Capacitor ESR	c_r	36 mΩ
Inductance	L	25 μH
Inductor resistance	l_r	0.12 Ω
Switching frequency	f_S	200 kHz
Minimum load resistance (step-down)	$r_{L(min)}$	6 Ω
Minimum load resistance (step-up)	$r_{L(min)}$	18 Ω
Maximum load resistance (step-down)	$r_{L(max)}$	120 Ω
Maximum load resistance (step-up)	$r_{L(max)}$	700 Ω
Desired output voltage (step-down)	V_{od}	12 V
Desired output voltage (step-up)	V_{od}	36 V

7.4.3   Buck-Boost Converter

Table 7.7 shows the specifications of the buck-boost converter used in this
section. The converter is to be operated for both step-down and step-up conversions.

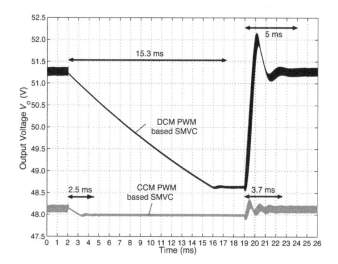

(a) CCM operation; r_L alternating between 48 Ω and 192 Ω

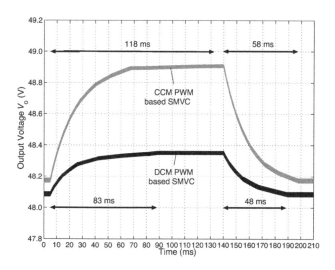

(b) DCM operation; r_L alternating between 480 Ω and 800 Ω

FIGURE 7.8
Waveforms of output voltage v_o of the SMVC boost converter with the CCM PWM-based and the DCM PWM-based controllers operating in (a) CCM and (b) DCM operations.

7.4.3.1   Steady-State Performance

Figures 7.9 and 7.10 show respectively the steady-state behavior of the buck-boost converter with the DCM PWM-based controller for step-down and step-up conversions. Similar to the case of the boost converter, the peak amplitude of the ramp signal also changes with load resistance (7.54)).

Figure 7.11(a) and 7.11(b) show respectively the variation of the DC output voltage for different operating load resistances for step-down and step-up conversions. When the buck-boost converter is operated in CCM and step-down mode, i.e., at load resistance $r_L < 22.5$ Ω (critical resistance), the CCM PWM-based controller achieves a better load regulation with a -0.130 V deviation in v_o (i.e., $\frac{dv_o}{dr_L} = -7.88$ mV/Ω) than the DCM PWM-based controller, which has a -1.109 V deviation in v_o (i.e., $\frac{dv_o}{dr_L} = -67.21$ mV/Ω), for the load range 6 $\Omega \leq r_L \leq 22.5$ Ω. For DCM operation, i.e., at load resistance $r_L > 22.5$ Ω (critical resistance), the CCM PWM-based controller *also* achieves a better load regulation with a 0.207 V deviation in v_o (i.e., $\frac{dv_o}{dr_L} = 2.12$ mV/Ω) than the DCM PWM-based controller, which has a -0.430 V deviation in v_o (i.e., $\frac{dv_o}{dr_L} = -4.41$ mV/Ω), for the load range 22.5 $\Omega \leq r_L \leq 120$ Ω. This indicates that the PWM-based controller derived from the DCM design approach has poorer load regulation property than the CCM design approach in the case of step-down conversion of the buck-boost converter operating in DCM.

When the buck-boost converter is operated in CCM and step-up mode, i.e., at load resistance $r_L < 63$ Ω (critical resistance), the CCM PWM-based controller achieves a better load regulation with a -0.129 V deviation in v_o (i.e., $\frac{dv_o}{dr_L} = -2.87$ mV/Ω) than the DCM PWM-based controller, which has a -1.601 V deviation in v_o (i.e., $\frac{dv_o}{dr_L} = -35.58$ mV/Ω), for the load range 18 $\Omega \leq r_L \leq 63$ Ω. For DCM operation, i.e., at load resistance $r_L > 63$ Ω (critical resistance), the DCM PWM-based controller achieves a better load regulation with a -0.615 V deviation in v_o (i.e., $\frac{dv_o}{dr_L} = -0.97$ mV/Ω) than the CCM PWM-based controller, which has a 1.580 V deviation in v_o (i.e., $\frac{dv_o}{dr_L} = 2.48$ mV/Ω), for the load range 63 $\Omega \leq r_L \leq 700$ Ω. Hence, unlike in the case of step-down conversion, the DCM PWM-based controller performs better in regulating the load than the CCM PWM-based controller when the buck-boost converter operates in DCM for step-up conversion.

7.4.3.2   Transient Performance

The dynamic performance of the controllers as applied to the buck-boost converter in step-down mode is studied using a load resistance that alternates between 6 Ω and 15 Ω for CCM operation, and between 30 Ω and 120 Ω for DCM operation. Figures 7.12(a) and 7.12(b) show the output voltage ripple waveforms. When applied to the buck-boost converter in CCM operation, the CCM PWM-based controller achieves shorter settling times (1.2 ms and 2.2 ms) than the DCM PWM-based controller (2.2 ms and 2.6 ms). Conversely,

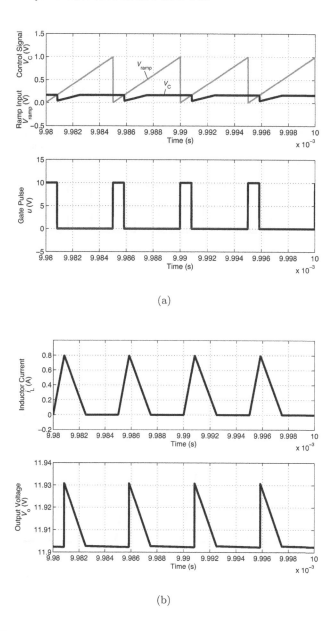

FIGURE 7.9
Waveforms of (a) control signal v_c, input ramp v_{ramp}, gate pulse u, and (b) inductor current i_L and output voltage ripple v_o of the PWM-based SMVC buck-boost converter in DCM step-down operation at nominal condition.

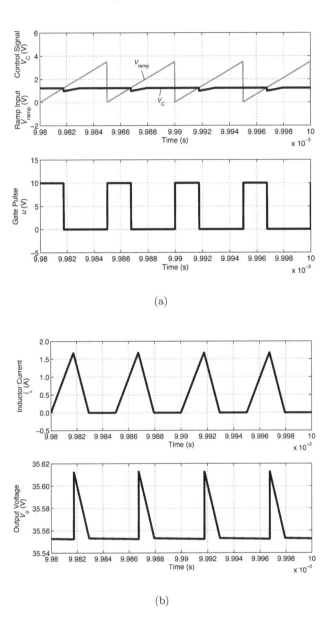

FIGURE 7.10
Waveforms of (a) control signal v_c, input ramp v_{ramp}, gate pulse u, and (b) inductor current i_L and output voltage ripple v_o of the PWM-based SMVC buck-boost converter DCM step-up operation at nominal condition.

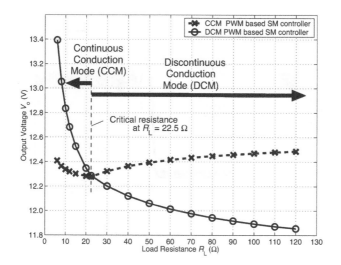

(a) Step-down conversion

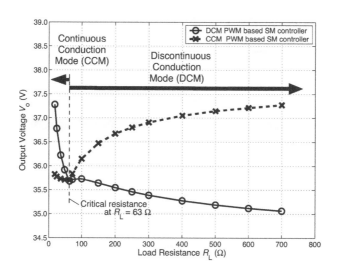

(b) Step-up conversion

FIGURE 7.11
DC output voltage $\overline{V}_o$ versus load resistance r_L of SMVC buck-boost converter designed for (a) step-down conversion and (b) step-up conversion, using the CCM PWM-based and the DCM PWM-based controllers.

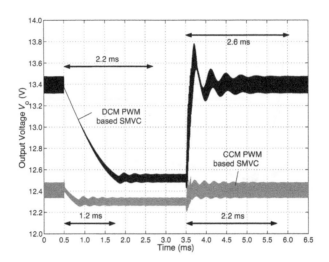

(a) CCM operation; r_L alternating between 6 Ω and 15 Ω

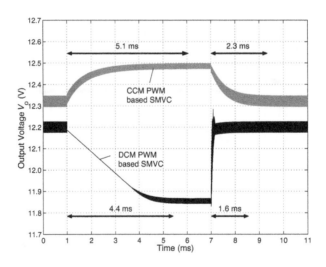

(b) DCM operation; r_L alternating between 30 Ω and 120 Ω

FIGURE 7.12
Waveforms of output voltage v_o of the SMVC buck-boost converter designed for step-down conversion using the CCM PWM-based and the DCM PWM-based controllers operating in (a) CCM and (b) DCM operations.

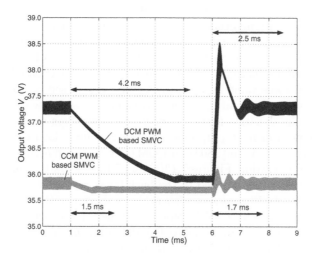

(a) CCM operation; r_L alternating between 18 Ω and 180 Ω

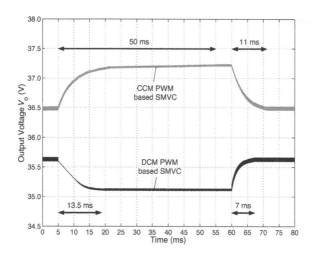

(b) DCM operation; r_L alternating between 150 Ω and 600 Ω

FIGURE 7.13

Waveforms of output voltage v_o of the SMVC buck-boost converter designed for step-up conversion using the CCM PWM-based and the DCM PWM-based controllers operating in (a) CCM and (b) DCM operations.

for DCM operation, the DCM PWM-based controller achieves shorter settling times (4.4 ms and 1.6 ms) than the CCM PWM-based controller (5.1 ms and 2.3 ms). It should be pointed out that even though the load regulation performance is poorer for the case of step-down conversion and DCM operation, the transient performance is better with the DCM PWM-based controller. Hence, there is some value in using the DCM PWM-based controller for the buck-boost converter operating in DCM designed for step-down conversion.

Similarly, the dynamic performance of the controllers as applied to the buck-boost converter in step-up mode is studied using a load resistance that alternates between 18 Ω and 180 Ω for CCM operation, and between 150 Ω and 600 Ω for DCM operation. Figures 7.13(a) and 7.13(b) show the output voltage ripple waveforms. When applied to the buck-boost converter in CCM operation, the CCM PWM-based controller achieves shorter settling times (1.5 ms and 1.7 ms) than the DCM PWM-based controller (4.2 ms and 2.5 ms). Conversely, for DCM operation, the DCM PWM-based controller achieves shorter settling times (13.5 ms and 7.0 ms) than the CCM PWM-based controller (50.0 ms and 11.0 ms).

7.5 Other Application of DCM SM Control: Hybrid Dual-Operating-Mode Controllers

7.5.1 Background

In practice, converters are always designed for operation in either CCM or DCM, and never in both modes. A converter that is designed to operate in both modes will require a controller that is capable of providing the appropriate control according to the operating mode that the converter assumes at different times and under different conditions. Existing PWM current or voltage-mode controllers are incapable of achieving that. They can only be designed for either one of the two operating modes. A severe degradation in control performance is experienced when the converter enters into an operating mode that its controller is not designed for. This explains why only single-mode (functional in either CCM or DCM only) converters have been normally adopted in practice.

Nevertheless, the use of converters that can operate in both modes does have its advantages. Firstly, such dual-mode converters are can be used over a wider range of operating condition than conventional single-mode converters. Secondly, under the same power rating, the inductor size(s) of a dual-mode converter can be made smaller than that of the single-mode CCM converter since the latter must always be designed to have an inductance larger than the critical inductance while the former does not have such a concern. Both these factors lead to more economical use of the power devices and magnetic

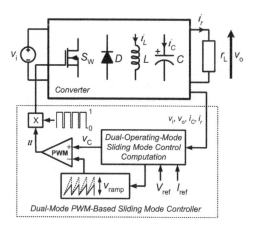

FIGURE 7.14

Overview of the dual-operating-mode PWM-based SM-controlled DC–DC converter.

components in the dual-mode converter than the single-mode converter. However, before the dual-mode converters can be adopted in actual applications, controllers that can facilitate this kind of operation must first be developed.

Interestingly, we find that the dual-operating-mode controllers can be easily realized using the principle of SM control. Common features are found in the control architecture between the class of PWM-based SM controllers for CCM converters and the class of PWM-based SM controllers for DCM converters. The main difference lies in the multiplication factor related to the output load. This structural resemblance provides the possibility of developing a hybrid type of SM controllers (combination of both classes of controllers) that is functional in dual-mode converters.

7.5.2 Architecture

With the difference being only the equivalent switching component, it is possible to merge the two classes of controllers into a single hybrid class of controllers that cater for both modes of operation, by introducing a control logic that switches between the two sets of control equations according to the converter's operating mode. Figure 7.14 shows an overview of the so-called dual-operating-mode (DOM) PWM-based SM controller.

Designed for converters that operate in both CCM and DCM, the DOM PWM-based SM controllers require the feedback of the four state variables, i.e., i_C, i_r, v_i, and v_o. However, a main difference between the DOM PWM-based SM controllers and the PWM-based SM controllers is the use of the load current reference term I_{ref}. This is an identifier for distinguishing the state of the operating mode in which the converter is engaged. A load current below the reference value, i.e., $i_r < I_{ref}$, indicates that the converter is in DCM, and a load current above the reference value, i.e., $i_r > I_{ref}$, indicates that the converter is in CCM. Typically, this reference value can be easily determined from the converter's specifications.

With the identifier incorporated, the DOM controllers adopt the control equations of the controllers for DCM converters (Table 7.4) with an additional control logic of the following expression:

$$u_V = \begin{cases} \text{as in Table 7.4} & \text{when } i_r < I_{ref} \\ 1 & \text{when } i_r \geq I_{ref} \end{cases}. \qquad (7.55)$$

Under this control scheme, the DOM controllers will operate as PWM-based SM voltage controllers for DCM converters with control equations given in Table 7.4 when the load current is detected to be less than the reference current, i.e., $i_r < I_{ref}$, and they will operate as PWM-based SM voltage controllers for CCM converters with control equations given in Table 6.3 when the load current is detected to be equal or above the reference current, i.e., $i_r \geq I_{ref}$.

Finally, the method of selecting the control parameters K_{p1} and K_{p2} are based on satisfying the existence and stability conditions for SM operation as derived in the cases of CCM and DCM operations.

7.5.3 Simulation Results and Discussions

The DOM PWM-based SM voltage controllers are compared against the the previously discussed PWM-based SM voltage controllers. Results and discussions are provided for the buck and boost converters.

7.5.3.1 Buck Converter

Table 7.5 shows the specifications of the buck converter used in the simulation. The controller is designed with a bandwidth of $f_{BW} = 20$ kHz for the maximum load current (i.e., $r_{L(min)}$). Choosing $V_{ref} = 2.5$ V and $\beta = 0.208$ V/V, the sliding coefficients are calculated as $\frac{\alpha_1}{\alpha_2} = 251327$ s^{-1} and $\frac{\alpha_3}{\alpha_2} = 15791367040$ s^{-2} (see Chapter 8 for the method of calculation). The control parameters are then determined as $K_{p1} = \beta L \left(\frac{\alpha_1}{\alpha_2} - \frac{1}{r_{L(min)}C} \right) = 1.15$ Hs^{-1}, $K_{p2} = \frac{\alpha_3}{\alpha_2} LC = 52.11$ V/V, and $\sqrt{\frac{2L}{T}} = 2.97$ $\Omega^{-0.5}$. Therefore, the control equation of the DOM PWM-based SM voltage controller is

$$\begin{cases} v_c = -1.15i_C + 52.11 \left(V_{ref} - \beta v_o \right) + 0.208 v_o u_V \\ \hat{v}_{ramp} = 0.208 v_i \\ u_V = \begin{cases} 2.97 \sqrt{\frac{v_i}{v_o(v_i - v_o)}} \sqrt{i_r} & \text{when } i_r < I_{ref} \\ 1 & \text{when } i_r \geq I_{ref} \end{cases} \end{cases}. \qquad (7.56)$$

Figures 7.15(a)–7.15(c) show the plots of the DC output voltage against

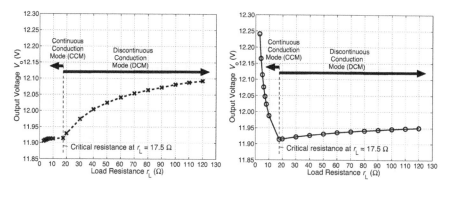

(a) Controller designed for CCM (b) Controller designed for DCM

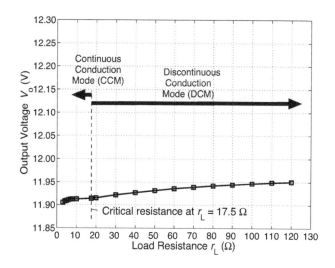

(c) Hybrid dual-operating-mode SM controller

FIGURE 7.15
DC output voltage v_o against load resistance r_L of the buck converter with
the (a) PWM-based SM voltage controller designed for CCM buck converter;
(b) PWM-based SM voltage controller designed for DCM buck converter; and
(c) hybrid dual-operating-mode PWM-based SM voltage controller.

the different operating load resistances of the buck converter under the control
of the various PWM-based SM voltage controllers. It can be observed from
Figs. 7.15(a) and 7.15(b) that for operation in CCM, i.e., at load resistance
$r_L < 17.5\ \Omega$, the controller designed for the CCM converter has a better load

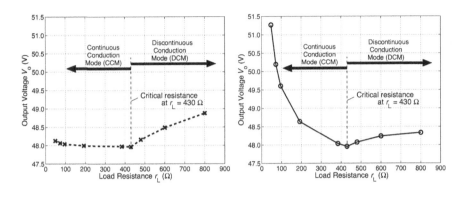

(a) Controller designed for CCM   (b) Controller designed for DCM

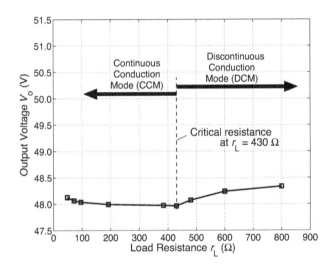

(c) Hybrid dual-operating-mode SM controller

FIGURE 7.16
DC output voltage v_o against load resistance r_L of the boost converter with the (a) PWM-based SM voltage controller designed for CCM boost converter; (b) PWM-based SM voltage controller designed for DCM boost converter; and (c) hybrid dual-operating-mode PWM-based SM voltage controller.

regulation than the controller designed for DCM converter, and conversely, for operation in DCM, i.e., $r_L > 17.5\ \Omega$, the controller designed for the DCM converter has a better load regulation. Figure 7.15(c) shows the simulation result of the buck converter under the control of the hybrid DOM controller. It is shown that the DOM controller is capable of providing excellent regulation to the buck converter in both operating modes.

7.5.3.2 Boost Converter

Table 7.6 shows the specifications of the boost converter used in the simulation. The controller is designed with a bandwidth of $f_{BW} = 2$ kHz for the maximum load current (i.e., minimum load resistance $r_{L(min)}$). Choosing $V_{ref} = 8$ V and $\beta = 0.167$ V/V, the sliding coefficients are calculated as $\frac{\alpha_1}{\alpha_2} = 25132.7$ s^{-1} and $\frac{\alpha_3}{\alpha_2} = 157913670$ s^{-2}. The control parameters are then determined as $K_{p1} = 0.209$ Hs^{-1}, $K_{p2} = 7.896$ V/V, and $\sqrt{\frac{2L}{T}} = 4.47$ $\Omega^{-0.5}$. Therefore, the control equation is

$$\begin{cases} v_c = -0.209 i_C + 7.896\,(V_{ref} - \beta v_o) \\ \qquad\qquad\qquad + 0.167\,(v_o - v_i)\,u_V \\ \hat{v}_{ramp} = 0.167\,(v_o - v_i)\,u_V \\ u_V = \begin{cases} 4.47\frac{v_o}{v_i}\sqrt{\frac{1}{v_o - v_i}}\sqrt{i_r} & \text{when } i_r < I_{ref} \\ 1 & \text{when } i_r \ge I_{ref} \end{cases} \end{cases} \tag{7.57}$$

Figures 7.16(a)–7.16(c) show the plots of the DC output voltage against the different operating load resistances of the boost converter under the control of the various PWM-based SM voltage controllers. Similar to the case of the buck converter, the controller designed for the CCM converter has a better load regulation than the controller designed for DCM converter when the boost converter operates in CCM. This is also true for the case of DCM operation. Figure 7.16(c) shows the simulation result of the boost converter under the control of the hybrid DOM controller. This illustrates again the capability of the DOM controller in providing excellent regulation to converters in both operating modes.

8

Design and Implementation of PWM-Based Sliding Mode Controllers for Power Converters

CONTENTS

8.1   Introduction

In the previous two chapters, a general approach for designing the pulse-width modulation (PWM)-based SM voltage controllers for buck, boost, and buck-boost converters has been expounded. In this chapter, we continue to discuss the design and implementation of the PWM-based SM voltage controllers, with emphasis on practical circuit design methods. It must be stressed that the design of these controllers at the circuit level involves a different set of engineering considerations. It is not an easy task especially if one aims to achieve a high level of accuracy using the simplest possible hardware for the circuit design. In addition to the circuit design, a practical approach to the design and selection of the sliding coefficients of the controller is introduced. This approach, which is based on Ackermann's formula [1], permits the control design to be carried out systematically.

As an illustration of how such controllers can be achieved, several design examples, namely the PWM-based SMVC buck converters, the PWM-based SMVC boost converters, the PWM-based SM current controlled boost converters, the PWM-based SM controlled Ćuk converter, and the PWM-based SM controller for switched-capacitor converters, are provided. The controllers for these converters are designed in analog form, which are suitable for industrial applications. Experiments are carried out to validate the theoretical design. Nevertheless, while the subject of discussion is limited to the few types of controllers, the same procedure can be employed for the design of other PWM-based SM power converters.

8.2   PWM-based SM Voltage Controller for Buck Converters

This section covers the implementation and design of the PWM-based SM voltage controller for buck converters. For consistency, the same proportional-integral-derivative (PID) SM voltage controller discussed in the previous chapters is adopted in the experimental work. The discussion is focused on the continuous conduction mode operation. The schematic diagram of the PWM-based SM voltage controlled buck converter is given in Fig. 6.2(a).

8.2.1 Mathematical Model

For ease of reading, we restate the existence condition of the buck converter given in Table 6.2, i.e.,

$$0 < -\beta L \left(\frac{\alpha_1}{\alpha_2} - \frac{1}{r_L C} \right) i_C + LC \frac{\alpha_3}{\alpha_2} \left(V_{\text{ref}} - \beta v_o \right) + \beta v_o < \beta v_i \qquad (8.1)$$

and the control equations in Table 6.3, i.e.,

$$v_c = -\beta L \left(\frac{\alpha_1}{\alpha_2} - \frac{1}{r_L C} \right) i_C + \frac{\alpha_3}{\alpha_2} LC \left(V_{\text{ref}} - \beta v_o \right) + \beta v_o \qquad (8.2)$$

and

$$\hat{v}_{\text{ramp}} = \beta v_i, \qquad (8.3)$$

which are directly applicable for the implementation of the controller.

On the other hand, the inequalities in (8.1) give the conditions for existence and therefore provide a range of employable sliding coefficients that will ensure that the converter stays in SM operation when its state trajectory is near the sliding surface.

8.2.2 Existence Condition with Design Parameters Consideration

In practice, it would be useful to tighten the design constraints of the existence condition by absorbing the actual operating parameters into the inequality. This can be done by decomposing (8.1) into two inequalities and considering them as individual cases with respect to the direction of the capacitor current flow. Since in practice $\frac{\alpha_1}{\alpha_2} > \frac{1}{r_L C}$, the left inequality of (8.1) is implied by

$$0 < LC \frac{\alpha_3}{\alpha_2} \left(V_{\text{ref}} - \beta v_o \right) - \beta L \left(\frac{\alpha_1}{\alpha_2} - \frac{1}{r_L C} \right) \left| \hat{i}_C \right| + \beta v_o, \qquad (8.4)$$

which can be rearranged to give

$$\frac{\alpha_1}{\alpha_2} < \frac{\beta v_o + LC \frac{\alpha_3}{\alpha_2} \left(V_{\text{ref}} - \beta v_o \right)}{\beta L \left| \hat{i}_C \right|} + \frac{1}{r_L C}, \qquad (8.5)$$

and the right inequality of (8.1) is implied by

$$LC \frac{\alpha_3}{\alpha_2} \left(V_{\text{ref}} - \beta v_o \right) + \beta L \left(\frac{\alpha_1}{\alpha_2} - \frac{1}{r_L C} \right) \left| \hat{i}_C \right| + \beta v_o < \beta v_i, \qquad (8.6)$$

which can be rearranged to give

$$\frac{\alpha_1}{\alpha_2} < \frac{\beta v_i}{\beta L \left| \hat{i}_C \right|} - \frac{\beta v_o + LC \frac{\alpha_3}{\alpha_2} \left(V_{\text{ref}} - \beta v_o \right)}{\beta L \left| \hat{i}_C \right|} + \frac{1}{r_L C}, \qquad (8.7)$$

where $\hat{i_C}$ is the peak amplitude of the bidirectional capacitor current flow. Next, equations (8.5) and (8.7) can be recombined and further tightened by considering the range of input and loading conditions of the converter to give

$$\frac{\alpha_1}{\alpha_2} < \frac{v_o + LC\frac{\alpha_3}{\alpha_2}\left(\frac{V_{\text{ref}}}{\beta} - v_o\right)}{L\left|\hat{i_C}\right|} + \frac{1}{R_{\text{L(max)}}C}$$

$$\text{for } V_{\text{i(min)}} \geq 2\left[v_o + LC\frac{\alpha_3}{\alpha_2}\left(\frac{V_{\text{ref}}}{\beta} - v_o\right)\right]$$

$$\frac{\alpha_1}{\alpha_2} < \frac{V_{\text{i(min)}}}{L\left|\hat{i_C}\right|} - \frac{v_o + LC\frac{\alpha_3}{\alpha_2}\left(\frac{V_{\text{ref}}}{\beta} - v_o\right)}{L\left|\hat{i_C}\right|} + \frac{1}{R_{\text{L(max)}}C}$$

$$\text{for } V_{\text{i(min)}} < 2\left[v_o + LC\frac{\alpha_3}{\alpha_2}\left(\frac{V_{\text{ref}}}{\beta} - v_o\right)\right] \qquad (8.8)$$

where $R_{\text{L(max)}}$ is the maximum load resistance and $V_{\text{i(min)}}$ is the minimum input voltage, which the converter is designed for. Additionally, the peak capacitor current $\left|\hat{i_C}\right|$ is the maximum inductor current ripple during steady-state operation.

Theoretically, one may assume that at steady-state operation, the actual output voltage v_o is ideally a pure DC waveform whose magnitude is equal to the desired output voltage $V_{\text{od}} \equiv V_{\text{ref}}/\beta$. However, this is not true in practice. Due to the limitation of finite switching frequency and imperfect feedback loop, there will always be some steady-state DC error between v_o and V_{od}, even in the presence of an error-reducing integral controller (i.e., PI, PID). It is important to take this error into consideration for the design of the controller since the factor $LC\frac{\alpha_3}{\alpha_2}\left(\frac{V_{\text{ref}}}{\beta} - v_o\right)$ is relatively large in comparison to v_o.

Now, assuming that

(a) for controllers with integral control function, the difference between v_o and V_{od} is small, and when optimally designed, is normally limited to a range of within ± 1 % of V_{od}[1];

and in our particular controller arrangement where the voltage error is denoted as $V_{\text{ref}} - \beta v_o$,

(b) the DC average of v_o is always lower than V_{od} for PWM-based SMVC converters;

(c) the term $LC\frac{\alpha_3}{\alpha_2}$ is always positive,

we can rewrite the existence condition (8.8) for the PWM-based SMVC con-

---

[1]This is a conservative value to be adopted for output voltage accuracy. Many switching regulators available have errors of less than ± 1 %.

verter as

$$\frac{\alpha_1}{\alpha_2} < \frac{\left(0.99 + 0.01 \, LC\frac{\alpha_3}{\alpha_2}\right) V_{od}}{L\left|\hat{i_C}\right|} + \frac{1}{R_{L(max)}C}$$

$$\text{for } V_{i(min)} \geq \left(1.98 + 0.02 \, LC\frac{\alpha_3}{\alpha_2}\right) V_{od}$$

$$\frac{\alpha_1}{\alpha_2} < \frac{V_{i(min)}}{L\left|\hat{i_C}\right|} - \frac{\left(0.99 + 0.01 \, LC\frac{\alpha_3}{\alpha_2}\right) V_{od}}{L\left|\hat{i_C}\right|} + \frac{1}{R_{L(max)}C}$$

$$\text{for } V_{i(min)} < \left(1.98 + 0.02 \, LC\frac{\alpha_3}{\alpha_2}\right) V_{od} \qquad (8.9)$$

by substituting $v_o = 0.99 \, V_{od}$ as appropriate. Thus, the control parameters α_1, α_2, and α_3 are now bounded by inequalities more stringently than in (8.8).

8.2.3 Selection of Sliding Coefficients

Clearly, inequalities (8.9) provide only the general information for the existence of SM, but give no details about the selection of the parameters. Nor does it give any information relating the sliding coefficients to the converter performance. For this purpose, we employ the Ackermann's formula for designing static controllers [1]. This basically concerns with the selection of sliding coefficients based on the desired dynamic properties. In this way, the stability condition of the system is automatically satisfied. This is a direct approach of assuring stability, whereby the same objective of making the *eigenvalues* of the *Jacobian* of the system in SM operation to contain negative real parts is achieved.

In this example, the equation relating the sliding coefficients to the dynamic response of the converter during SM operation can be easily found by substituting $S = 0$ into (6.10), i.e.,

$$\alpha_1 x_1 + \alpha_2 \frac{dx_1}{dt} + \alpha_3 \int x_1 dt = 0. \qquad (8.10)$$

Rearranging the time differentiation of (8.10) into a standard second-order system form, we have

$$\frac{d^2 x_1}{dt^2} + 2\zeta\omega_n \frac{dx_1}{dt} + \omega_n^2 x_1 = 0 \qquad (8.11)$$

where $\omega_n = \sqrt{\frac{\alpha_3}{\alpha_2}}$ is the undamped natural frequency and $\zeta = \frac{\alpha_1}{2\sqrt{\alpha_2\alpha_3}}$ is the damping coefficient. These equations can be rearranged as

$$\frac{\alpha_3}{\alpha_2} = \frac{1}{4\zeta^2} \left(\frac{\alpha_1}{\alpha_2}\right)^2. \qquad (8.12)$$

Recall that there are three possible types of responses in a linear second-order system: under-damped ($0 \leq \zeta < 1$), critically-damped ($\zeta = 1$), and over-damped ($\zeta > 1$).

(a) For the under-damped response converters, when SM operation is assumed, the output voltage error is expressed as

$$x_1(t) = A_1 \frac{e^{-\zeta \omega_n t}}{\sqrt{1 - \zeta^2}} \cos \left(\omega_n \sqrt{1 - \zeta^2} t - \psi \right), \qquad \text{for } t \geq 0 \qquad (8.13)$$

where A_1 is determined by the initial condition of the system and phase angle $\psi = \arctan \frac{\zeta}{\sqrt{1-\zeta^2}}$. Since the rate of decay in the response is determined by $\zeta \omega_n$, the time constant for the decay is

$$\tau = \frac{1}{\zeta \omega_n} = 2 \frac{\alpha_2}{\alpha_1}. \qquad (8.14)$$

Assuming that it takes 5τ s (1% criteria) for the system to settle to its steady state, the sliding coefficient ratio of α_1 and α_2 with respect to the settling time T_s is

$$\frac{\alpha_1}{\alpha_2} = \frac{10}{T_s}. \qquad (8.15)$$

This allows the desired settling time to be set by tuning $\frac{\alpha_1}{\alpha_2}$. In addition, for such systems, the peak oscillating overshoot is dependent only on the damping coefficient and is expressed as

$$\zeta = \sqrt{\frac{\left[\ln \left(\frac{M_p}{100} \right) \right]^2}{\pi^2 + \left[\ln \left(\frac{M_p}{100} \right) \right]^2}} \qquad (8.16)$$

where M_p is the percentage of the peak overshoot. Hence, for some desired overshoot, the damping coefficient ζ can be calculated using (8.16). Substitution of (8.15) into (8.12) gives $\frac{\alpha_3}{\alpha_2}$.

$$\frac{\alpha_3}{\alpha_2} = \frac{25}{\zeta^2 T_s^2}. \qquad (8.17)$$

Hence, the required ratio for $\frac{\alpha_3}{\alpha_2}$ can be obtained by using the calculated ζ and desired T_s.

(b) For the critically-damped response converters, when SM operation is assumed, the output voltage error is expressed as

$$x_1(t) = (A_1 + A_2 t)e^{-\omega_n t}, \qquad \text{for } t \geq 0 \qquad (8.18)$$

where A_1 and A_2 are determined by the initial conditions of the system.

Since the rate of decay in the response is determined by ω_n, the time constant for the decay is

$$\tau = \frac{1}{\omega_n} = \sqrt{\frac{\alpha_2}{\alpha_3}}. \tag{8.19}$$

With $\zeta = 1$, substitution of (8.12) into (8.19), we have

$$\tau = 2\frac{\alpha_2}{\alpha_1}. \tag{8.20}$$

Assuming a 5τ s (1% criteria) settling time T_s, the required sliding coefficient ratios are

$$\frac{\alpha_1}{\alpha_2} = \frac{10}{T_s} \quad \text{and} \quad \frac{\alpha_3}{\alpha_2} = \frac{25}{T_s^2}. \tag{8.21}$$

(c) For the over-damped response converters, when SM operation is assumed, the output voltage error is expressed as

$$x_1(t) = A_1 \left[\frac{-\zeta + \sqrt{\zeta^2 - 1}}{2\sqrt{\zeta^2 - 1}} e^{\left(-\zeta - \sqrt{\zeta^2 - 1}\right)\omega_n t} \right.$$
$$\left. - \frac{-\zeta - \sqrt{\zeta^2 - 1}}{2\sqrt{\zeta^2 - 1}} e^{\left(-\zeta + \sqrt{\zeta^2 + 1}\right)\omega_n t} \right] \text{ for } t \geq 0, \tag{8.22}$$

where A_1 is determined by the initial condition of the system. Here, the rate of decay in the response is affected by two factors, $\left(\zeta - \sqrt{\zeta^2 - 1}\right)\omega_n$ and $\left(\zeta + \sqrt{\zeta^2 - 1}\right)\omega_n$. It is sufficient to consider the slower of the two, i.e., $\left(\zeta - \sqrt{\zeta^2 - 1}\right)\omega_n$, for the computation of the settling time. For this component, the time constant for the decay is

$$\tau = \frac{1}{\left(\zeta - \sqrt{\zeta^2 - 1}\right)\omega_n} = \frac{2}{1 - \sqrt{1 - \frac{1}{\zeta^2}}} \frac{\alpha_2}{\alpha_1}. \tag{8.23}$$

Assuming a 5τ s (1% criteria) settling time T_s, the following equations can be formulated from (8.23) and (8.12):

$$\frac{\alpha_1}{\alpha_2} = \frac{10}{\left(1 - \sqrt{1 - \frac{1}{\zeta^2}}\right) T_s} \quad \text{and} \quad \frac{\alpha_3}{\alpha_2} = \frac{25}{\left(\zeta - \sqrt{\zeta^2 - 1}\right) T_s^2}. \tag{8.24}$$

Hence, for any desired settling time and damping coefficient, where $\zeta > 1$, the ratios $\frac{\alpha_1}{\alpha_2}$ and $\frac{\alpha_3}{\alpha_2}$ can be obtained from (8.24).

Note that the 5τ s (1% criteria) settling time assumed in the derivation is the time taken for the controller to complete the *SM operation phase*, i.e., the time taken for the state trajectory S to track from any point on the sliding surface to the steady-state equilibrium. This time period does not include the *reaching phase*, which requires a relatively small, but finite, time period for S to move from an arbitrary position (depending on the magnitude of the load or line disturbance and the capacity of the energy storage elements) to the sliding surface. Theoretically, the total time taken to complete both the SM operation phase and the reaching phase is equivalent to the settling time T_s for the converter. However, in practice, since the time taken to complete the SM operation phase is usually much longer than that of the reaching phase, it is sufficient to consider only the time duration of the SM operation phase as the settling time. Furthermore, it should also be noted that the aforementioned settling time description is only valid for an ideal SM controller with infinite switching frequency. For a finite-switching-frequency SM controller, i.e., PWM-based SM controller, the settling time will be slightly longer than the ideal case due to the non-ideality of the tracking motion.

Thus, the design of the sliding coefficients is now dependent on the desired settling time of the response and the type and amount of damping required, in conjunction with the existence condition of the respective PWM-based controllers. It is worth mentioning that the design equations in (8.15), (8.17), (8.21) and (8.24) are applicable to all types of power converters.

8.2.4 Implementation of Controller

Figure 8.1 shows the schematic diagram of the PWM-based SMVC buck converter. The controller architecture is based on (8.2) and (8.3). Briefly, the design of this controller can be summarized as follows: selection of the desired frequency response's bandwidth, calculation of the corresponding sliding coefficients; verification of compliance with the existence condition; and derivation of the control equations by substituting the calculated parameters into (8.2) and (8.3). We will next go through the design steps in detail.

8.2.4.1 Design Procedure

Step 1: Given a desired bandwidth, the control parameters can be calculated using (8.15) and (8.17), or (8.21), or (8.24). Substitute these parameters and the converter's parameters into (8.9) to verify that the existence condition is met.

Step 2: Given a reference voltage V_{ref}, β is calculated using the expression:

$$\beta = \frac{V_{\text{ref}}}{V_{\text{od}}}. \tag{8.25}$$

Also, R_1 and R_2 are related by

$$R_2 = \frac{\beta}{1 - \beta} R_1. \tag{8.26}$$

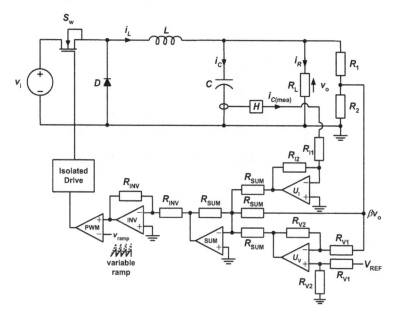

FIGURE 8.1
Schematic diagram of the proposed PWM-based SMVC buck converter.

Step 3: From (8.2), the gain required for the amplification of the signal $(V_{\text{ref}} - \beta v_o)$ is $\frac{\alpha_3}{\alpha_2} LC$. Hence, with known converter's parameters, R_{V1} and R_{V2} can be determined using

$$R_{V2} = \left(\frac{\alpha_3}{\alpha_2} LC \right) R_{V1}. \qquad (8.27)$$

Step 4: Setting the current sensing gain H at a value such that the measured capacitor current $i_{\text{C(mea)}}$ is equal to the actual capacitor current i_C, and considering that the controller is designed for maximum load current (i.e., minimum load resistance $R_{\text{L(min)}}$), the gain required for the amplification of the signal $i_{\text{C(mea)}}$ is $\beta L \left(\frac{1}{R_{\text{L(min)}} C} - \frac{\alpha_1}{\alpha_2} \right)$. Hence, with known converter's parameters, R_{I1} and R_{I2} can be determined using

$$R_{I2} = \beta L \left(\frac{1}{R_{\text{L(min)}} C} - \frac{\alpha_1}{\alpha_2} \right) R_{I1}. \qquad (8.28)$$

Step 5: Figure 8.2 shows the schematic diagram of the adaptive feedforward variable ramp generator adopted in our controller design. According to the figure, the rate of change of voltage in the capacitor C_{ramp} is

$$\frac{dv_{\text{ramp}}}{dt} = \frac{v_{\text{i}}}{R_{\text{ramp}} C_{\text{ramp}}}. \qquad (8.29)$$

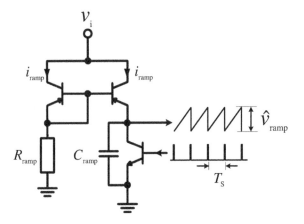

FIGURE 8.2
Schematic diagram of the adaptive feedforward ramp signal generator.

For one complete ramp cycle of T duration, v_{ramp} is linearly varied from 0 V to $\hat{v}_{\text{ramp}}$ V. Hence,

$$\hat{v}_{\text{ramp}} = \frac{v_{\text{i}}}{R_{\text{ramp}}C_{\text{ramp}}}T. \tag{8.30}$$

Since $\hat{v}_{\text{ramp}} = \beta v_{\text{i}}$ and $T = \frac{1}{f_{\text{s}}}$, the above equation can be arranged as

$$R_{\text{ramp}} = \frac{1}{\beta C_{\text{ramp}}f_{\text{s}}}. \tag{8.31}$$

8.2.4.2 Parameters of Controllers

The design approach described above and the analog realization for the PWM-based SMVC buck converter are verified through simulations[2] and experiments. The specifications of the converter is given in Table 8.1.

The converter is designed to operate in the continuous conduction mode for $v_{\text{i}} = 16$ V to 30 V and $i_R = 0.5$ A to 4 A. The calculated critical inductance is $L_{\text{crit}} = 36$ μH. The minimum required capacitance is $C_{\text{min}} = 9$ μF. The maximum allowable peak-to-peak ripple voltage is 50 mV.

The controller is designed for critically-damped response for two different bandwidths: at one-twentieth and at one-tenth of the switching frequency f_{S}, i.e., $f_{\text{BW}} = 10$ kHz (i.e., first-order response time constant $\tau_{10 \text{ kHz}} = 15.915$ μs and settling time $T_{\text{s}(10 \text{ kHz})} = 79.575$ μs) and $f_{\text{BW}} = 20$ kHz (i.e., $\tau_{20 \text{ kHz}} = 7.956$ μs settling time $T_{\text{s}(20 \text{ kHz})} = 39.780$ μs). Assuming that the controller is

---

[2]The simulation is performed using Matlab/Simulink. The step size taken for all simulations is 10 ns.

TABLE 8.1

Specifications of buck converter.

Description	Parameter	Nominal Value
Input voltage	v_i	24 V
Capacitance	C	150 μF
Capacitor ESR	c_r	21 mΩ
Inductance	L	100 μH
Inductor resistance	l_r	0.12 Ω
Switching frequency	f_S	200 kHz
Minimum load resistance	$R_{L(min)}$	3 Ω
Maximum load resistance	$R_{L(max)}$	24 Ω
Desired output voltage	V_{od}	12 V

designed for maximum load current (i.e., minimum load resistance $R_{L(min)}$), the reference voltage is chosen as $V_{ref} = 2.5$ V, the ratio of the voltage divider network $\beta = 0.208$ V/V, and the current sensing ratio is set at $H = 1$ A/A so that $i_C = i_{C(mea)}$, the parameters of the controllers can be calculated as follows.

8.2.4.3 Parameters of 10 kHz Bandwidth Controller

Substituting $T_{s(10\ kHz)} = 79.575\ \mu$s into (8.21), the sliding coefficients can be determined as $\frac{\alpha_1}{\alpha_2} = 125667.6\ \text{s}^{-1}$ and $\frac{\alpha_3}{\alpha_2} = 3948086999\ \text{s}^{-2}$. Note that for the full-load condition, we can readily verify that $\frac{\alpha_1}{\alpha_2} \gg \frac{1}{R_{L(min)}C}$. Finally, the control parameters are determined as $K_{p1} = \beta L \left(\frac{\alpha_1}{\alpha_2} - \frac{1}{R_{L(min)}C} \right) = 2.572\ \text{Hs}^{-1}$ and $K_{p2} = \frac{\alpha_3}{\alpha_2} LC = 59.218$ V/V. Therefore, for the 10 kHz bandwidth controller, the control equation calculated from the design equations is

$$v_c = -2.572 i_C + 0.208 v_o + 59.218 \left(V_{ref} - \beta v_o \right). \tag{8.32}$$

The actual values of the components used in the simulation and experiment are

$$R_1 = 950\ \Omega, \quad R_2 = 250\ \Omega, \quad R_{I1} = 2.16\ \text{k}\Omega, \quad R_{I2} = 5.6\ \text{k}\Omega, \tag{8.33}$$
$$R_{V1} = 2.2\ \text{k}\Omega, \quad R_{V2} = 130\ \text{k}\Omega, \quad R_{INV} = 10\ \text{k}\Omega, \quad R_{SUM} = 10\ \text{k}\Omega.$$

Hence, the actual control equation used in our simulation and experimental study is

$$v_c = -2.593 i_C + 0.208 v_o + 59.100 \left(V_{ref} - \beta v_o \right). \tag{8.34}$$

8.2.4.4 Parameters of 20 kHz Bandwidth Controller

Similarly, substituting $T_{s(20\ kHz)} = 39.780\ \mu$s into (8.21) gives the sliding coefficients as $\frac{\alpha_1}{\alpha_2} = 251327.41\ \text{s}^{-1}$ and $\frac{\alpha_3}{\alpha_2} = 15791367040\ \text{s}^{-2}$. Note that designing for the full-load condition, we require that $\frac{\alpha_1}{\alpha_2} \gg \frac{1}{R_{L(min)}C}$, i.e., $251327.41 \text{ s}^{-1} \gg 2222.2 \text{ s}^{-1}$. Finally, the control parameters are determined as $K_{p1} = \beta L \left(\frac{\alpha_1}{\alpha_2} - \frac{1}{R_{L(min)}C} \right) = 5.190 \text{ Hs}^{-1}$ and $K_{p2} = \frac{\alpha_3}{\alpha_2} LC = 236.875 \text{ V/V}$. Therefore, for the 20 kHz bandwidth controller, the ideal control equation calculated from the design equations is

$$v_c = -5.190 i_C + 0.208 v_o + 236.875 \left(V_{ref} - \beta v_o \right). \tag{8.35}$$

The values of the components used in the controller for both the simulation and experiment are

$$R_1 = 950 \text{ } \Omega, \quad R_2 = 250 \text{ } \Omega, \quad R_{I1} = 2.16 \text{ k}\Omega, \quad R_{I2} = 11.2 \text{ k}\Omega, \tag{8.36}$$
$$R_{V1} = 3.6 \text{ k}\Omega, \quad R_{V2} = 910 \text{ k}\Omega, \quad R_{INV} = 10 \text{ k}\Omega, \quad R_{SUM} = 10 \text{ k}\Omega.$$

Hence, the implemented control equation is

$$v_c = -5.185 i_C + 0.208 v_o + 252.780 \left(V_{ref} - \beta v_o \right). \tag{8.37}$$

8.2.4.5   Parameters of Adaptive FeedForward Ramp Generator

The input ramp signal for both the controllers is a sawtooth signal that varies from 0 V to its peak amplitude $\hat{v}_{ramp} = 0.208 \, v_i$ V at a constant frequency $f_S = 200$ kHz. Setting $C_{ramp} = 330$ pF, and by using (8.31), R_{ramp} is calculated as 72.7 kΩ. A 75 kΩ resistor is chosen. Figure 8.3 shows the full schematic diagram of the experimental prototype.

8.2.5   Results and Discussions

8.2.5.1   Steady-State Performance

Figure 8.4 shows the experimental waveforms during steady-state operation for the SMVC converter with the 20 kHz bandwidth controller operating at full load (i.e., $r_L = 3 \, \Omega$). Figure 8.5 shows the corresponding set of experimental waveforms for the SMVC converter with the 10 kHz bandwidth controller operating at full load (i.e., $r_L = 3 \, \Omega$). Due to the higher magnitude of the sliding coefficients, v_c of the 20 kHz bandwidth controller has a higher peak-to-peak value than v_c of the 10 kHz bandwidth controller.

8.2.5.2   Load Variation Analysis

Figure 8.6 shows a plot of the measured DC output voltage versus the operating load resistance. At full load operation (i.e., $r_L = 3 \, \Omega$), the converter employing the 20 kHz bandwidth controller has a steady-state DC output voltage $\overline{V}_o$ of 11.661 V, which corresponds to a -2.825% deviation from V_{od}. The plot also shows that even though $\overline{V}_o$ increases with r_L, $\overline{V}_o$ is always less than V_{od}. This is in line with the discussion given in Section 8.2.2. part (b)

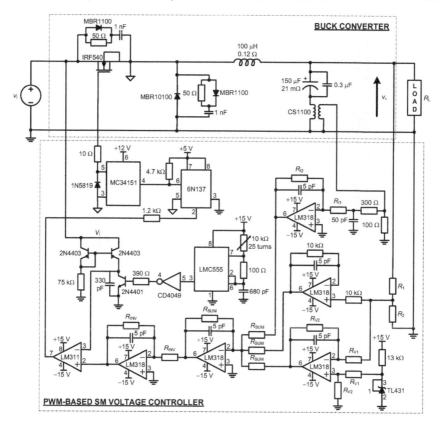

FIGURE 8.3
Full schematic diagram of the PWM-based SMVC buck converter prototype.

that the output voltage of the PWM-based SMVC system is always falling below the desired voltage. Furthermore, it also shows that the converter has satisfactory load regulation, having only a 0.151 V deviation in $\overline{V}_o$ for the entire load range of $3\ \Omega \leq r_L \leq 24\ \Omega$, i.e., the load regulation is only 1.29% of $\overline{V}_{o(\text{full load})}$ from full load to minimum load.

For the converter employing the 10 kHz bandwidth controller, the steady-state DC output voltage at full load operation is 11.633 V, which corresponds to a -3.058% deviation from V_{od}. For the entire load range of $3\ \Omega \leq r_L \leq 24\ \Omega$, $\overline{V}_o$ has a deviation of 0.189 V, i.e., the load regulation is 1.62% of $\overline{V}_{o(\text{full load})}$ from full load to minimum load. Thus, it can be concluded that the 20 kHz bandwidth controller has a better load variation property than the 10 kHz bandwidth controller.

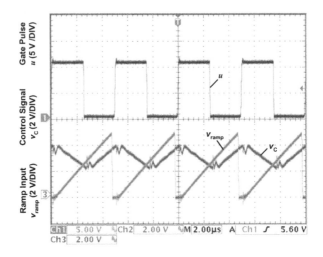

(a) v_c, v_{ramp}, and u

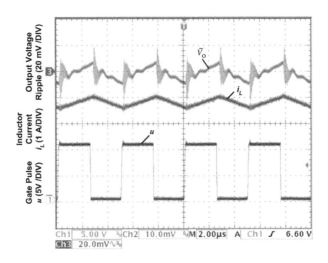

(b) u, i_L, and $\tilde{v}_o$

FIGURE 8.4
Experimental waveforms of (a) control signal v_c, input ramp v_{ramp}, and generated gate pulse u and (b) waveforms of gate pulse u, and the corresponding inductor current i_L and output voltage ripple $\tilde{v}_o$ (right), for the SMVC converter with the 20 kHz bandwidth controller operating at load resistance $r_L = 3\ \Omega$.

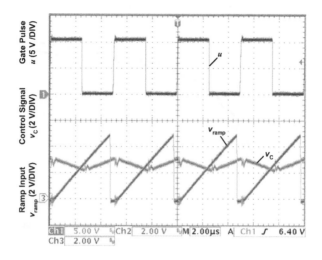

(a) v_c, v_{ramp}, and u

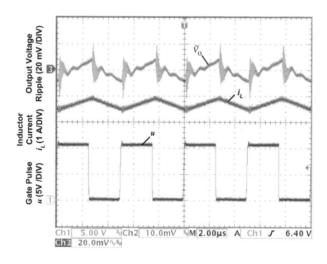

(b) u, i_L, and $\widetilde{v}_o$

FIGURE 8.5

Experimental waveforms of (a) control signal v_c, input ramp v_{ramp}, and generated gate pulse u and (b) waveforms of gate pulse u, and the corresponding inductor current i_L and output voltage ripple $\widetilde{v}_o$ (right), for the SMVC converter with the 10 kHz bandwidth controller operating at load resistance $r_L = 3\ \Omega$.

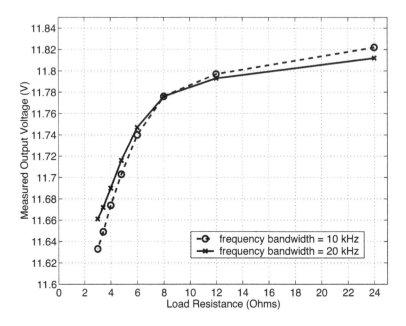

FIGURE 8.6
Measured DC output voltage $\overline{V}_o$ versus load resistance r_L for SMVC buck converter with the 10 kHz and 20 kHz bandwidth controllers.

8.2.5.3 Line Variation Analysis

Figure 8.7 shows the experimental waveforms of the SMVC buck converter with a minimum and maximum input voltage of $v_i = 16$ V and $v_i = 30$ V. As shown in the figure, the controller operates effectively for both operating conditions.

Furthermore, to investigate the effectiveness of the adaptive feedforward control property, experiments are performed for both the cases with and without the use of an adaptive ramp signal .

Figures 8.8(a)–8.8(d) and Figs. 8.9(a)–8.9(d) show, respectively, the experimental waveforms of the converter with the minimum and maximum input voltage. The waveforms with and without the adaptive feedforward control can be differentiated in terms of their ramp signals ($\hat{v}_{ramp} = 5.00$ V in Figs. 8.8(a), 8.8(c), 8.9(a), and 8.9(c); $\hat{v}_{ramp} = 3.33$ V in Figs. 8.8(b) and 8.8(d); and $\hat{v}_{ramp} = 6.25$ V in Figs. 8.9(b) and 8.9(d)). The other difference between the adaptive control and non-adaptive control is manifested in the waveforms of the control signal v_c. When $v_i = 16$ V, with a lower $\hat{v}_{ramp}$, the adaptive feedforward controller attempts to accommodate a constant duty ratio $d = \frac{v_o}{v_i}$ by reducing the magnitude of v_c as shown in Fig. 8.8. This effectively tightens the voltage regulation relative to the controller without the adaptive feedforward control. In contrast, when the converter operates at $v_i = 30$ V, the magnitude

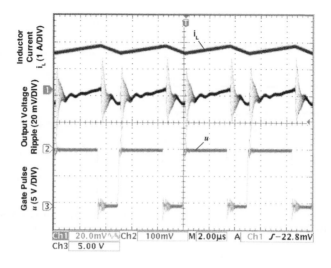

(a) $v_i = 16$ V

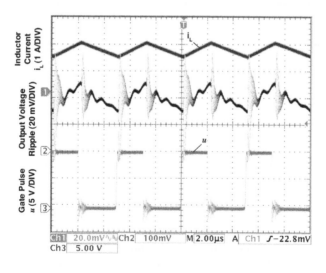

(b) $v_i = 30$ V

FIGURE 8.7
Experimental waveforms of gate pulse u, and the corresponding inductor current i_L and output voltage ripple $\tilde{v}_o$, for the SMVC converter with the 10 kHz bandwidth controller operating at input voltage (a) $v_i = 16$ V and (b) $v_i = 30$ V (right), at load resistance $r_L = 3$ Ω.

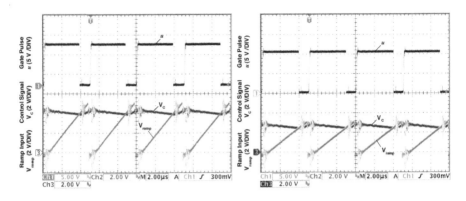

(a) 10 kHz bandwidth controller without adaptive feedforward control

(b) 10 kHz bandwidth controller with adaptive feedforward control

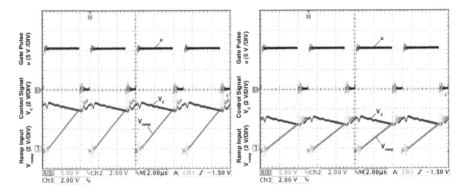

(c) 20 kHz bandwidth controller without adaptive feedforward control

(d) 20 kHz bandwidth controller with adaptive feedforward control

FIGURE 8.8
Experimental waveforms of control signal v_c, input ramp v_{ramp}, and generated gate pulse u for the SMVC converter with both the 10 kHz and 20 kHz bandwidth controllers, with and without the adaptive feedforward control property, operating at input voltage $v_i = 16$ V and load resistance $r_L = 3$ Ω.

of v_c is automatically raised by the adaptive feedforward control (compare Figs. 8.9(a) and 8.9(c) with Figs. 8.9(b) and 8.9(d)). This, on the other hand, loosens the voltage regulation. Surprisingly, such control actions of tightening and loosening the voltage regulation as input voltage varies, are actually the inherited virtues of the adaptive feedforward control. This is evident from Fig. 8.10 where it is observed that the line regulation from minimum to maximum input voltage is corrected from 1.38% of $\overline{V}_{o(v_i=24 \text{ V})}$ (10 kHz bandwidth con-

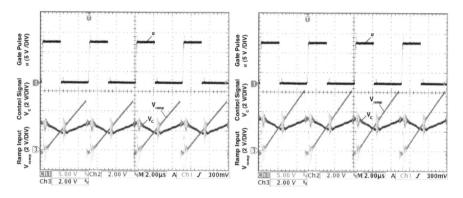

(a) 10 kHz bandwidth controller without adaptive feedforward control

(b) 10 kHz bandwidth controller with adaptive feedforward control

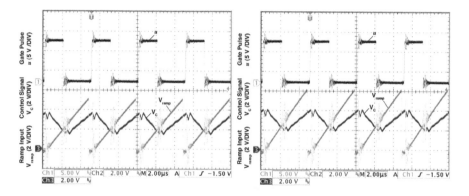

(c) 20 kHz bandwidth controller without adaptive feedforward control

(d) 20 kHz bandwidth controller with adaptive feedforward control

FIGURE 8.9
Experimental waveforms of control signal v_c, input ramp v_{ramp}, and generated gate pulse u for the SMVC converter with both the 10 kHz and 20 kHz bandwidth controllers, with and without the adaptive feedforward control scheme, operating at input voltage $v_i = 30$ V and load resistance $r_L = 3$ Ω.

troller without adaptive feedforward control) and 0.43% of $\overline{V}_{o(v_i=24\ V)}$ (20 kHz bandwidth controller without adaptive feedforward control), to a perfect regulation of 0% for both controllers with the adaptive feedforward control.

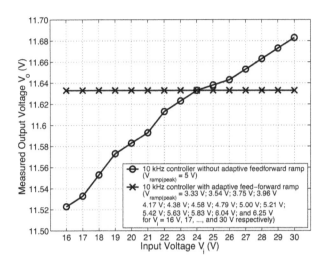

(a) 10 kHz controllers

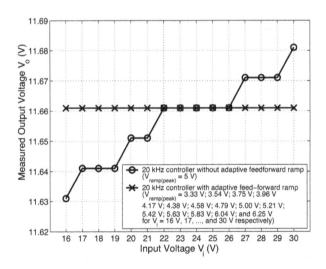

(b) 20 kHz controllers

FIGURE 8.10

Measured DC output voltage $\overline{V}_o$ versus input voltage v_i for SMVC buck converter with the (a) 10 kHz and (b) 20 kHz bandwidth controllers, with and without the adaptive feedforward control property.

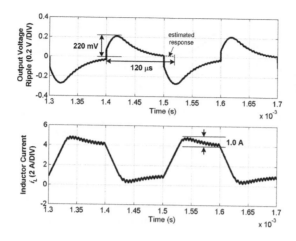

(a) 10 kHz controllers

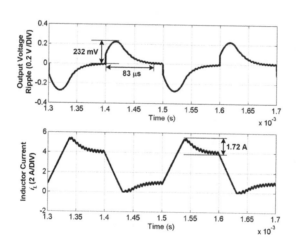

(b) 20 kHz controllers

FIGURE 8.11
Simulated waveforms of output voltage ripple $\tilde{v}_o$ and inductor current i_L of the SMVC converter with (a) the 10 kHz bandwidth controller and (b) the 20 kHz bandwidth controller, with the load being stepped between $r_L = 3\ \Omega$ and $r_L = 12\ \Omega$ at 5 Hz.

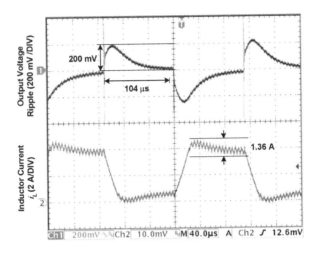

(a) 10 kHz controllers

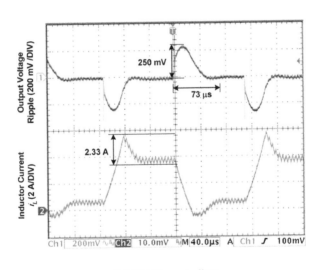

(b) 20 kHz controllers

FIGURE 8.12

Experimental waveforms of output voltage ripple $\tilde{v}_o$ and inductor current i_L of the SMVC converter with (a) the 10 kHz bandwidth controller and (b) the 20 kHz bandwidth controller, with the load being stepped between $r_L = 3\ \Omega$ and $r_L = 12\ \Omega$ at 5 Hz.

8.2.5.4 Dynamic Performance

The dynamic performance of the controllers is studied using a load resistance that alternates between a quarter load (12 Ω) and the full load (3 Ω) at a constant frequency of 5 kHz. Figures 8.11 and 8.12 show, respectively, the simulated and experimental output voltage ripple (top) and inductor current (bottom) waveforms of the converter for the 10 kHz bandwidth controller (left) and the 20 kHz bandwidth controller (right). As illustrated in Fig. 8.11, the simulated output voltage has an overshoot ripple of 220 mV (1.83% of V_{od}) and a steady-state settling time of 120 μs for the 10 kHz bandwidth controller, and an overshoot ripple of 232 mV (1.93% of V_{od}) and a steady-state settling time of 83 μs for the 20 kHz bandwidth controller, during the load transients. As shown in Fig. 8.12, the output voltage has an overshoot ripple of 200 mV (1.67% of V_{od}) and a steady-state settling time of 104 μs for the 10 kHz bandwidth controller, and an overshoot ripple of 250 mV (2.08% of V_{od}) and a steady-state settling time of 73 μs for the 20 kHz bandwidth controller, during the load transients. Furthermore, consistent with a critically-damped response, there is no ringing or oscillation during the transient phase.

However, it should also be mentioned that there are some slight disagreements between the experimental and simulation results in terms of the overshoot ripple magnitudes and the settling times. These are mainly due to the small discrepancy of the model used in the simulation program, and the parameters' deviation of the actual experimental circuits from the simulation program due to the variation of the actual components used in the setup. Additionally, it should be clarified that the settling time measured in this case is the total settling time taken to complete both the SM operation phase and the reaching phase of the control process. Hence, the steady-state settling time exceeds 5τ s, where $\tau = \tau_{10\ kHz}$ for the 10 kHz bandwidth controller and $\tau = \tau_{20\ kHz}$ for the 20 kHz bandwidth controller.

8.2.5.5   A Comparison with Classical PWM Voltage-Mode Controller

The dynamic behavior of the PWM-based SMVC buck converter is compared with that of the conventional type of PWM voltage-mode controlled buck converter. In the experiment, the former employs a 20 kHz bandwidth PWM-based SM controller and the latter employs a PID PWM voltage-mode controller that is optimally tuned to operate at a step load change that alternates between $r_L = 3\ \Omega$ and $r_L = 12\ \Omega$. Figures 8.13(a)–8.13(f) show the experimental waveforms for both converters operating at 5 kHz step load change.

Under the conventional PWM voltage-mode control, the system shows different dynamic responses for different operating conditions. Specifically, the response becomes more oscillatory at lower currents. As shown in Figs. 8.13(a), 8.13(c), and 8.13(e), the output voltage waveforms show different oscillatory responses. Such variation in the dynamic response is expected since the PWM controller is designed for a specific operating conditions.

On the other hand, with the PWM-based SM controller, the dynamic behavior of the system is basically unaffected (i.e., critically damped) by the change in operating conditions, even when it enters momentarily into discon-

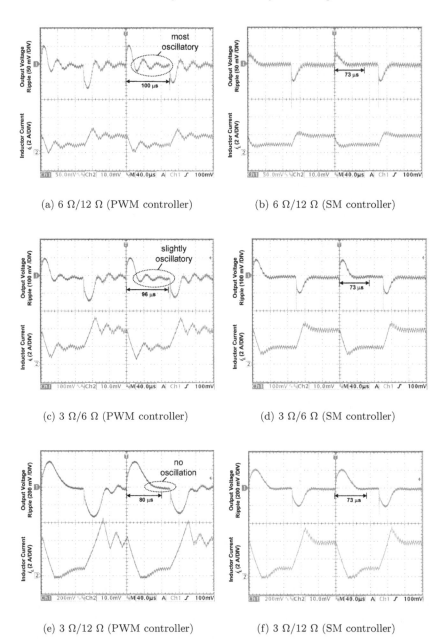

(a) 6 Ω/12 Ω (PWM controller)   (b) 6 Ω/12 Ω (SM controller)

(c) 3 Ω/6 Ω (PWM controller)   (d) 3 Ω/6 Ω (SM controller)

(e) 3 Ω/12 Ω (PWM controller)   (f) 3 Ω/12 Ω (SM controller)

FIGURE 8.13

$\tilde{v}_o$ and i_L waveforms of the buck converter with PWM voltage controller (a), (c), (e) and PWM-based SM controller (b), (d), (f), at 5 kHz step load change.

tinuous conduction mode and experiences a change in converter's description. This demonstrates the strength of the SM controller in terms of robustness against changes in the operating conditions. In addition, the example illustrates a major difference between a large-signal controlled system (SM) and a small-signal controlled system (PWM). Specifically, the former is expected to produce a similar response for all operating conditions, while the latter aims to address a specific operating condition.

8.3 PWM-Based SM Voltage Controller for Boost Converters

This section covers the practical design and implementation of the PWM-based SMVC boost converter. Similar to the buck converter example, the same PID SMVC boost converter operating in continuous conduction mode discussed in the previous chapters is used in the experimental study. Figure 6.2(b) shows the schematic diagram of the converter.

8.3.1 Mathematical Model

For convenience, we restate the existence condition of the boost converter given in Table 6.2, i.e.,

$$0 < \beta L \left(\frac{\alpha_1}{\alpha_2} - \frac{1}{r_{\mathrm{L}} C} \right) i_C - LC \frac{\alpha_3}{\alpha_2} \left(V_{\mathrm{ref}} - \beta v_{\mathrm{o}} \right) < \beta (v_{\mathrm{o}} - v_{\mathrm{i}}) \qquad (8.38)$$

and the derived control equations in Table 6.3, i.e.,

$$v_{\mathrm{c}} = -\beta L \left(\frac{\alpha_1}{\alpha_2} - \frac{1}{r_{\mathrm{L}} C} \right) i_C + LC \frac{\alpha_3}{\alpha_2} \left(V_{\mathrm{ref}} - \beta v_{\mathrm{o}} \right) + \beta \left(v_{\mathrm{o}} - v_{\mathrm{i}} \right) \qquad (8.39)$$

and

$$\hat{v}_{\mathrm{ramp}} = \beta \left(v_{\mathrm{o}} - v_{\mathrm{i}} \right) \qquad (8.40)$$

which are directly applicable for the implementation of the controller.

In this case, the inequality in (8.38) gives the conditions for existence, which will provide a range of employable sliding coefficients. Similar to the buck converter example, it is possible to tighten these design constraints by absorbing the actual operating parameters and the non-zero steady-state error into the inequality. The method of selecting the sliding coefficients has been mentioned in the previous section, and is omitted here. We will next address the implementation issues of this controller.

8.3.2 Implementation of Controller

This section discusses some practical issues concerning the hardware implementation of the controller.

8.3.2.1 Control Signal Computation

The computation of the control signal v_c in (8.39) can be performed using the simple gain amplification and summing functions. In our prototype, we realize the equation using only three analog gain amplifiers and a summer circuit (LM318). The parameters of these circuitries can be easily calculated given the values of L, C, r_L, and β, and using proper choices of α_1, α_2, and α_3.

8.3.2.2 Bandwidth of Ramp Voltage Generator

The generation of the ramp voltage signal requires the sampling of the instantaneous input and/or output voltages (see Table 6.3). Variation in the load causing variation in the output voltage or variation in the input voltage will affect the ramp voltage. Hence, to ensure the correct generation of the ramp voltage signal according to the instantaneous input and/or output voltages, the bandwidth of the ramp voltage generator must be greater than the bandwidths of the input and/or output voltage variations. Note that the continuous adjustment of the peak amplitude of the ramp signal according to variations of the input and/or output voltages is part of the controller's effort to maintain, respectively, the line and load regulation properties of the system.

For the boost converter, the peak amplitude of the variable ramp signal $\hat{V}_{ramp}$ is defined by (8.40). In our prototype, an assembly of current-mirror circuits (9015 and 9016) and a charging capacitor are employed to realize the ramp generation. Since the desired output voltage is normally constant, only the input voltage change is considered in the design. The frequency of the ramp signal is controlled by an impulse generator (LMC555 and CD4049).

8.3.2.3 Duty-Ratio Protection

The incorporation of the control and ramp signal circuitries into the pulse-width modulator (LM311) forms the basic architecture of the PWM-based SM controller. However, recalling that the boost-type converter cannot operate with a switching signal u that has a duty ratio $d = 1$, a small protective circuitry is required to ensure that the duty ratio of the controller's output is always less than 1. In our prototype, this is satisfied by multiplying the logic state u_{PWM} of the pulse-width modulator with the logic state u_{CLK} of the impulse generator using a logic AND IC chip (CD4081). By doing so, the maximum duty ratio of the controller is clamped by the duty ratio of the impulse generator.

TABLE 8.2

Specifications of boost converter.

Description	Parameter	Nominal Value
Input voltage	v_i	24 V
Capacitance	C	230 μF
Capacitor ESR	c_r	69 mΩ
Inductance	L	300 μH
Inductor resistance	l_r	0.14 Ω
Switching frequency	f_S	200 kHz
Minimum load resistance (full load)	$R_{L(min)}$	24 Ω
Maximum load resistance (ten-percent load)	$R_{L(max)}$	240 Ω
Desired output voltage	V_{od}	48 V

TABLE 8.3

Theoretical description of signals at different test locations of PWM-based SM controller.

Test Location	Description
1	$K_{p1}i_C$
2	$-K_{p2}\left(V_{ref} - \beta v_o\right)$
3	$-\beta\left(v_o - v_i\right)$
4	V_c
5	$\hat{V}_{ramp}$
6	u_{PWM}
7	u_{CLK}
8	$u = u_{PWM} \cdot u_{CLK}$

8.3.3 Experimental Prototype

The proposed analog implementation of the PWM-based SMVC boost converter is verified experimentally. The experimental prototype of the PWM-based SMVC boost converter is constructed as specified in Table 8.2.

The PWM-based SM controller is designed to give an under-damped response at a bandwidth of $\omega_n = 1.25$ krad/s, i.e., $\tau = 0.8$ ms and $T_s = 4.0$ ms, and damping coefficient $\zeta = 0.2$. From (8.15) and (8.17), the sliding coefficients are determined as $\frac{\alpha_1}{\alpha_2} = 2500$ s^{-1} and $\frac{\alpha_3}{\alpha_2} = 39130435$ s^{-2}. Note that for full-load condition, we can verify that $\frac{\alpha_1}{\alpha_2} \gg \frac{1}{r_{L(min)}C}$. Setting the reference voltage in the controller as $V_{ref} = 8$ V, the feedback divider ratio can be calculated as $\beta = \frac{V_{ref}}{V_{od}} = \frac{1}{6}$ V/V. Finally, control parameters are determined as $K_{p1} = \beta L\left(\frac{\alpha_1}{\alpha_2} - \frac{1}{r_L C}\right) = 0.12$ Hs^{-1} and $K_{p2} = \frac{\alpha_3}{\alpha_2}LC = 2.7$ V/V.

Figure 8.14 shows the schematic diagram of the proposed PWM-based SMVC boost converter. The numbered nodes on the diagram represent different test locations in the controller where waveforms are captured and analyzed. The theoretical expression of the signals are derived and shown in Table 8.3.

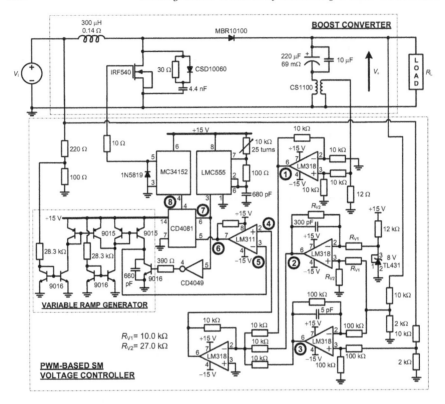

FIGURE 8.14
Schematic diagram of the proposed PWM-based SMVC boost converter.

8.3.4 Experimental Results and Discussions

This section presents a discussion on the results obtained from the experiments.

8.3.4.1 Measured Signals

Figure 8.15 shows the steady-state waveforms taken at the test locations when $r_L = 48\ \Omega$. The experimental results are consistent with the analysis (see Table 8.3).

8.3.4.2 Ensuring Duty-Ratio Protection

Figure 8.16 shows the waveforms measured at test locations 6, 7, and 8 when u_{PWM} (test point 6) has a duty ratio $d = 1$. It can be seen that the output signal u (test point 8) of the controller is clamped to $d < 1$ by the impulse signal u_{CLK} (test point 7). This validates the operation of the protection circuit discussed in the previous section.

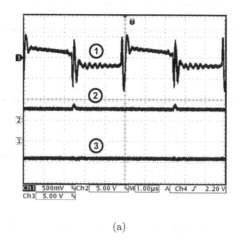

(a)

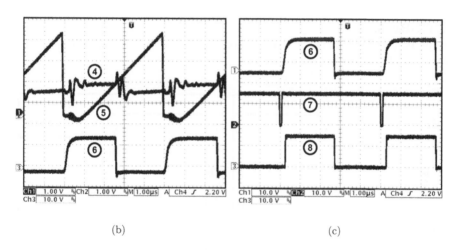

(b)   (c)

FIGURE 8.15
Experimental waveforms at the different test points when converter is operating at $r_L = 48\ \Omega$.

8.3.4.3 Testing of Variable Ramp Signal Generation

Figures 8.17(a) and 8.17(b) show the waveforms of ramp signal v_{ramp} corresponding respectively to a step-up and a step-down change in the input voltage. It can be observed that the time for the ramp signal to reach steady state after each step change between the minimum and maximum input voltage is about 10 μs. It takes at most two switching cycles for the ramp signal to adjust to any input voltage changes.

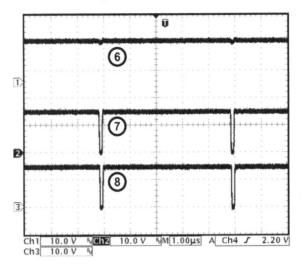

FIGURE 8.16
Experimental waveforms of test locations 6, 7, and 8 when u_{PWM} has a duty
ratio $d = 1$.

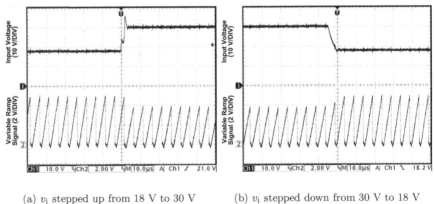

(a) v_{i} stepped up from 18 V to 30 V   (b) v_{i} stepped down from 30 V to 18 V

FIGURE 8.17
Waveforms of input voltage v_{i} and the corresponding ramp signal v_{ramp} of the
experimental prototype.

8.3.4.4   Control Signals at Different Input Voltage

Figures 8.18(a)–8.18(c) show the steady-state waveforms of the SM controller
for the boost converter with minimum, nominal, and maximum input voltages.
As explained earlier, the control signal v_{c} is produced from analog computation
involving the feedback signals v_{o}, v_{i}, and i_{c} using the expression described in

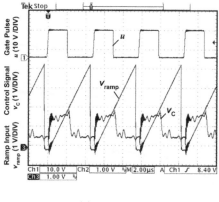

(a) $v_i = 20$ V

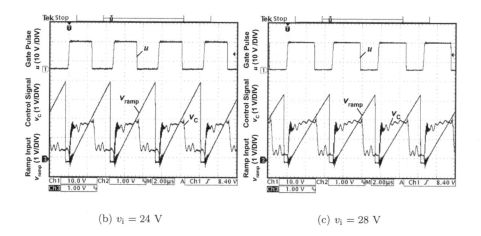

(b) $v_i = 24$ V   (c) $v_i = 28$ V

FIGURE 8.18
Steady-state waveforms of control signal v_c, input ramp v_{ramp}, and generated gate pulse u for the SMVC boost converter operating at full-load resistance $r_L = 24\ \Omega$.

(8.39). The ramp signal v_{ramp} is generated with peak amplitude as described in (8.40) using signals v_o and v_i through an analog variable ramp generator. Both v_c and v_{ramp} are fed into a pulse-width modulator to generate the gate pulse u, for controlling the switch of the boost converter. Clearly, the figures illustrate the difference in the ramp magnitude and the duty ratio with respect to different input voltages. Specifically, at $v_i = 20$ V, 24 V, and 28 V, the magnitude of v_{ramp} is about 4.67 V, 4.00 V, and 3.33 V, respectively, and the duty ratio d is roughly 0.56, 0.48, and 0.44, respectively.

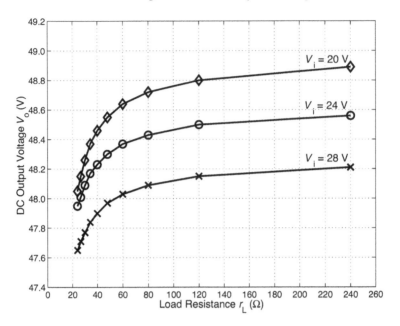

FIGURE 8.19

Plots of DC output voltage V_o against load resistance r_L for the PWM-based SMVC boost converter at minimum, nominal, and maximum input voltages.

TABLE 8.4

Load regulation property for minimum to maximum load. Output voltage at nominal operating condition $v_i = 24$ V and $r_L = 24\ \Omega$ is $v_{o(nominal)} = 47.95$ V.

Input Voltage	$\Delta v_o = v_{o(240\ \Omega)} - v_{o(24\ \Omega)}$	$\dfrac{\Delta v_o}{v_{o(nominal)}} \times 100\%$
$v_i = 20$ V	0.84 V	1.75% of $v_{o(nominal)}$
$v_i = 24$ V	0.61 V	1.27% of $v_{o(nominal)}$
$v_i = 28$ V	0.56 V	1.16% of $v_{o(nominal)}$

8.3.4.5 Regulation Performance

Figure 8.19 shows the measured DC output voltage versus the operating load resistance for three input voltage conditions, i.e., minimum, nominal, and maximum input voltages. Results of the load and line regulation properties are tabulated in Tables 8.4 and 8.5. According to Table 8.4, the maximum load-regulation error occurs at $v_i = 20$ V, with a deviation of 1.75% from $v_{o(nominal)}$. Similarly, it can be found from Table 8.5 that the maximum line-regulation error occurs at the minimum load $r_L = 240\ \Omega$, with a deviation of 1.42% from $v_{o(nominal)}$.

TABLE 8.5

Line regulation property for minimum to maximum input voltages. Output voltage at nominal operating condition $v_i = 24$ V and $r_L = 24$ Ω is $v_{o(nominal)} = 47.95$ V.

Load Condition	$\Delta v_o = v_{o(v_i=20\text{ V})} - v_{o(v_i=28\text{ V})}$	$\frac{\Delta v_o}{v_{o(nominal)}}$ x 100%
Minimum load (240 Ω)	0.68 V	1.42% of $v_{o(nominal)}$
Half load (48 Ω)	0.58 V	1.21% of $v_{o(nominal)}$
Full load (24 Ω)	0.40 V	0.83% of $v_{o(nominal)}$

8.3.4.6 Performance Comparison with Peak Current-Mode Controller

The dynamic behavior of the PWM-based controller is compared to that of the peak current-mode PWM controller (UC3843) that is optimally tuned to operate at a step load change from $r_L = 240$ Ω to $r_L = 24$ Ω for the input condition $v_i = 24$ V. Figures 8.20(a)–8.20(i) show the experimental waveforms of the peak current-mode controlled boost converter operating at a load resistance that alternates between $r_L = 24$ Ω and $r_L = 240$ Ω for various input voltages.

It can be seen that with the peak current-mode PWM controller, the dynamic behavior of the system differs for different operating conditions. Specifically, the response becomes less oscillatory at higher input voltages. Moreover, the dynamic behavior and transient settling time are also different for the various operating conditions. Specifically, at a lower step current change, i.e., 0.2 A to 1.0 A, the response of the system becomes critically-damped, instead of an optimally designed response which is slightly under-damped, as shown in Fig. 8.20(f). Furthermore, in the worst-case operating condition: $v_i = 20$ V and step output current change of 0.2 A to 2.0 A, the system has a settling time of 5.8 ms and a relatively high voltage ripple swing of 5.6 V (see Fig. 8.20(c)), which are much deviated from the optimally designed value of 2 ms and a voltage ripple swing of 2.6 V (see Fig. 8.20(f)). This is expected since the peak current-mode controller is designed using a linearized small-signal model that is only optimal for a specific operating condition. Thus, when the operating condition changes, the response varies.

On the other hand, with the PWM-based SM controller, the dynamic behavior of the output voltage ripple is basically similar (i.e., slightly under-damped) for all operating input and load conditions. This is illustrated in Figs. 8.21(a)–8.21(i), which show the experimental waveforms of the PWM-based SM controlled boost converter operating under the same set of operating conditions as the peak current-mode controlled boost converter.

Furthermore, the transient time, which is around 4.4 ms, is also independent of the direction and magnitude of the step load change and the operating input voltages. This is consistent with the design specification of the 1.25 krad/s bandwidth controller which is expected to have a settling time

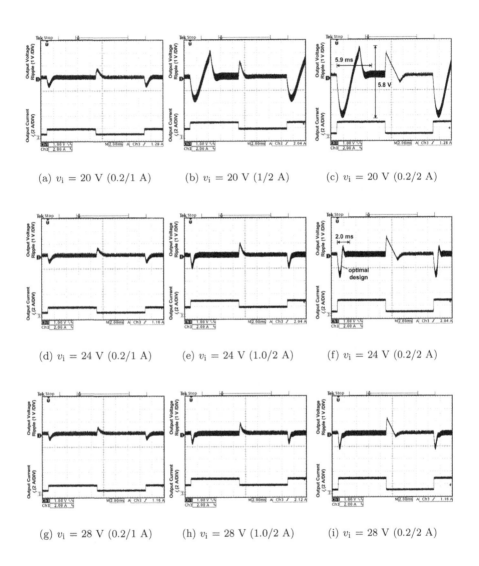

(a) $v_i = 20$ V (0.2/1 A)   (b) $v_i = 20$ V (1/2 A)   (c) $v_i = 20$ V (0.2/2 A)

(d) $v_i = 24$ V (0.2/1 A)   (e) $v_i = 24$ V (1.0/2 A)   (f) $v_i = 24$ V (0.2/2 A)

(g) $v_i = 28$ V (0.2/1 A)   (h) $v_i = 28$ V (1.0/2 A)   (i) $v_i = 28$ V (0.2/2 A)

FIGURE 8.20

Experimental waveforms of output voltage ripple $\tilde{v}_o$ and output current i_r of the boost converter with the peak current-mode controller operating at input voltage 20 V (minimum), 24 V (nominal), and 28 V (maximum), and alternating between load resistances 24 Ω (minimum), 48 Ω (half), and 240 Ω (maximum).

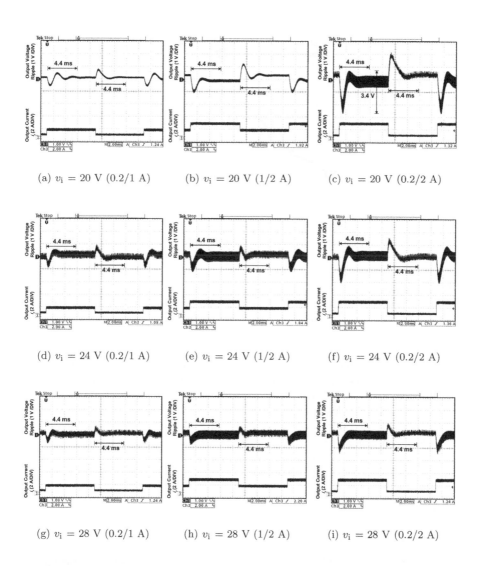

(a) $v_i = 20$ V (0.2/1 A)　　(b) $v_i = 20$ V (1/2 A)　　(c) $v_i = 20$ V (0.2/2 A)

(d) $v_i = 24$ V (0.2/1 A)　　(e) $v_i = 24$ V (1/2 A)　　(f) $v_i = 24$ V (0.2/2 A)

(g) $v_i = 28$ V (0.2/1 A)　　(h) $v_i = 28$ V (1/2 A)　　(i) $v_i = 28$ V (0.2/2 A)

FIGURE 8.21
Experimental waveforms of output voltage ripple $\tilde{v}_o$ and output current i_r of the boost converter with the 1.5 krad/s bandwidth PWM-based SM controller operating at input voltage 20 V (minimum), 24 V (nominal), and 28 V (maximum), and alternating between load resistance 24 Ω (minimum), 48 Ω (half), and 240 Ω (maximum).

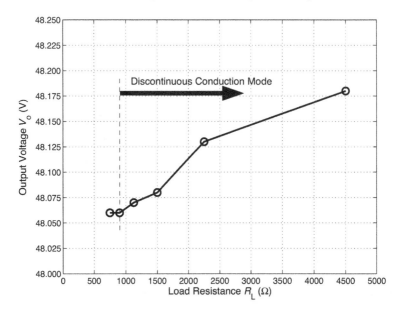

FIGURE 8.22
Graphs of DC output voltage $\overline{V}_o$ against load resistance r_L for the PWM-based SMVC boost converter in light load conditions.

of $5\tau = \frac{5}{1.25} = 4.0$ ms. Moreover, in the worst-case operating condition, i.e., $v_i = 20$ V and step output current change of 0.2 A to 2.0 A, the settling time is still around 4.4 ms and the voltage ripple swing is 3.4 V (see Fig. 8.21(c)). This performance is close to the optimally designed system, which has a voltage ripple swing of 3.0 V (see Fig. 8.21(f)). This illuminates again the strength of the SM controller in terms of robustness in the dynamic behavior against variation in operating conditions and uncertainties. It also reinforces our belief that a main advantage of the large-signal controller (SM) over the small-signal controller (PWM) is that it gives more consistent dynamic performances even when operating conditions are varying widely.

8.3.4.7   Operation in Discontinuous Conduction Mode

The PWM-based SMVC boost converter, which is designed for operation in the continuous conduction mode, is also tested in discontinuous conduction mode. Figure 8.22 shows the variation of the DC output voltage against different operating load resistances in the discontinuous conduction mode. The voltage regulation of the converter has a 0.12 V deviation (i.e., 0.25% of V_{od}) in v_o for the discontinuous conduction mode load range 900 $\Omega \leq r_L \leq 4500$ Ω, i.e., load regulation $\frac{dv_o}{dr_L}$ averages at 0.033 mV/Ω. Figures 8.23(a), 8.23(b), and 8.23(c) show respectively the behavior of the inductor current and the output voltage ripple when the converter is operating in discontinuous con-

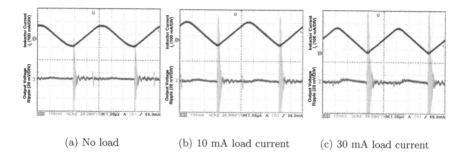

(a) No load   (b) 10 mA load current   (c) 30 mA load current

FIGURE 8.23
Waveforms of inductor current i_L and output voltage ripple $\widetilde{V_o}$ of the PWM-based SMVC boost converter operating in discontinuous conduction mode with (a) no operating load; (b) 10 mA load current; and (c) 30 mA load current.

duction mode at 0 A, 10 mA, and 30 mA output currents. It can be concluded from the results that the control is applicable to light load operations in the discontinuous conduction mode.

9

Sliding Mode Control with a Current Controlled Sliding Manifold

CONTENTS

9.1 Introduction

So far, in all previous chapters, we have been dealing only with the design and implementation of SM controllers of which the control variable used for constructing the sliding manifold is the output voltage. Using only the linear weighted summation of the voltage error, its time derivative, and time integral, the result is a linear sliding manifold of which the design can be easily accomplished through the use of the Ackermann's Formula for determining the control coefficients and a simple stability analysis of the sliding manifold.

In this and the following chapters, we will extend our discussion to cover SM control design for power converters with sliding manifold that is of a nonlinear composition of the current error and voltage error. The stability analysis of these controllers take a different approach from that of the linear sliding manifold. As an illustration of how such controllers can be designed and implemented, two design examples, namely the pulse-width modulation (PWM)-based SM current controlled boost converters will be discussed in this chapter and the PWM-based reduced-state SM controlled Ćuk converters in the next chapter. Similar to the SM voltage controllers, these controllers are implemented in analog form. Importantly, while the subject of discussion is limited to the two types of controllers, the same procedure can be employed for the design of SM controllers with other types of sliding manifolds.

9.2   The Need for Current-Mode Control in Boost-Type Converters

Boost-type converters operating in continuous conduction mode inherit the right-half-plane zero (RHPZ) characteristic in their duty-ratio-to-output-voltage transfer functions [33]. This typically makes the controlled dynamic response of the system sluggish, particularly when the control action is based solely on controlling the output voltage, i.e., voltage-mode control [33]. The presence of the RHPZ complicates the design and concurrently limits the bandwidth of the compensation network in the voltage-mode controller [72]. This is true for both linear and nonlinear types of voltage controllers.

A common solution to achieving fast dynamical response in RHPZ converter systems is to employ the current-mode control [47]. However, being a type of semi-linear control methodology, the current-mode control does not support converter applications over a very wide range of operating conditions. Dynamic responses with long settling times and large overshoots that can easily violate the specifications of the targeted application are observed when the operating conditions deviate sufficiently far from the desired points. This has spurred numerous investigations into the possible applications of various types of nonlinear controllers on power converters, all with the same objective of improving their controllability and performances for wide operating ranges [12, 13, 14, 15, 17, 24, 31, 52, 54, 64, 86, 93, 98, 87, 104, 108, 109, 110]. Despite their unique advantages, most of the controllers proposed above are impractical for power converters—either requiring complicated control circuitries [13, 14, 15, 24, 54, 64, 98, 87, 110], or being variable-frequency controllers [12, 17, 31, 52, 86, 104, 108, 109], or having a slow dynamical response [93].

The focus of this chapter is on demonstrating the application of SM control approach to designing current-mode control of power converters. Specifically, our discussion will be focused on the development of a fixed-frequency SM current controller for boost converters. The advantages of applying such a nonlinear controller to boost-type converters include 1) ability to operate under wide variation of operating conditions that cannot be satisfactorily controlled by conventional PWM current-mode controller, and 2) fast response that cannot be met by other SM or nonlinear voltage type of controllers. The main features are summarized as follows.

- Fast dynamical responses comparable to conventional current-mode controllers;

- Inherent robust features of SM control, but operating at a fixed frequency;

- Stability over a wide range of operating conditions;

- Small variation of the settling time over a wide range of operating conditions;

- Relatively low voltage overshoots (as compared to current-mode controllers) under large step changes over a wide range of operating conditions.

9.3 Sliding Mode Current Controller

The SM current controller employs both the output voltage error and the inductor current error as the controlled state variables. The incorporation of the output voltage error information allows the output voltage to be accurately regulated, whereas the inductor current error information allows the inductor current to follow closely the desired reference inductor current. As in the conventional current-mode controller, the monitoring and tracking of the inductor current reference is the key to achieving a fast dynamical response in RHPZ converter systems.

9.3.1 Generating a Suitable Reference Current Profile

Similar to the conventional PWM current-mode control, the instantaneous reference inductor current profile i_{ref} in the controller is generated using the amplified output voltage error, i.e.,

$$i_{\text{ref}} = K \left[V_{\text{ref}} - \beta v_{\text{o}} \right] \tag{9.1}$$

where V_{ref} and v_{o} denote the reference and instantaneous output voltages respectively; β denotes the feedback network ratio; and K is the amplified gain of the voltage error. A large value of K is chosen to improve dynamic response and to minimize the steady-state voltage error in the system. At steady-state operation, the reference current profile (i.e., the amplified steady-state voltage error) will be equivalent to the steady-state inductor current, i.e.,

$I_{\text{ref(ss)}} = I_{L(\text{ss})} = K\left[V_{\text{ref}} - \beta V_{\text{o(ss)}}\right]$. This would mean that $V_{\text{ref}} - \beta V_{\text{o(ss)}} \approx 0$ and not exactly zero. In such circumstances, the use of an additional integral action to generate the reference current profile is optional. It has insignificant influence on the steady-state regulation and can be ignored to simplify circuit implementation. Note that the main voltage regulation is performed by the SM control function, which has an integral control action to eliminate the output voltage regulation error. On the other hand, when a very tight regulation is required, the integral term can be included to give $i_{\text{ref}} = K_{\text{i}} \int\left[V_{\text{ref}} - \beta v_{\text{o}}\right] dt + K_{\text{p}}\left[V_{\text{ref}} - \beta v_{\text{o}}\right]$. The same mathematical treatment as provided in the following sections can be used to derive such a controller. In this discussion, the generation of reference current will be based on (9.1).

9.3.2   Sliding Surface

We denote S as the sliding surface, which can be chosen as a linear combination of the three state variables, i.e.,

$$S = \alpha_1 x_1 + \alpha_2 x_2 + \alpha_3 x_3 \qquad (9.2)$$

where α_1, α_2, and α_3 represent the sliding coefficients. The switching function u, which represents the logic state of power switch S_{W}, can be defined as $u = \frac{1}{2}\left(1 + \text{sign}(S)\right)$. Here, the controlled state variables are the *current error* x_1, the *voltage error* x_2, and the *integral of the current and the voltage errors* x_3, which are expressed as

$$\begin{cases} x_1 = i_{\text{ref}} - i_L \\ x_2 = V_{\text{ref}} - \beta v_{\text{o}} \\ x_3 = \int\left[x_1 + x_2\right] dt \\ \quad = \int\left(i_{\text{ref}} - i_L\right) dt + \int\left(V_{\text{ref}} - \beta v_{\text{o}}\right) dt \end{cases} \qquad (9.3)$$

where i_L denotes the instantaneous inductor current. Ideally, at an infinitely high switching frequency, only x_1 and x_2 are required in the SM current controller to ensure that both the output voltage and inductor current are tightly regulated, i.e., $v_{\text{o}} = V_{\text{ref}}$ and $i_L = i_{\text{ref}}$. However, in the case of finite-frequency or fixed-frequency SM controllers, the control is imperfect. Steady-state errors exist in both the output voltage error and the inductor current error such that $v_{\text{o}} \neq V_{\text{ref}}$ and $i_L \neq i_{\text{ref}}$. Therefore, an integral term of these errors x_3 has been introduced into the SM current controller as an additional controlled state variable to reduce these steady-state errors.

9.3.3   Dynamical Model of Controller/Converter System and Its Equivalent Control

Using the boost converter's behavioral model for continuous conduction mode and the time differentiation of (9.3), we obtain the dynamical model of the system as

$$
\begin{cases}
\dot{x}_1 = \dfrac{d[i_{\text{ref}} - i_L]}{dt} = -\dfrac{\beta K}{C} i_C - \dfrac{v_i - \bar{u}v_o}{L} \\[2mm]
\dot{x}_2 = \dfrac{d[V_{\text{ref}} - \beta v_o]}{dt} = -\dfrac{\beta}{C} i_C \\[2mm]
\dot{x}_3 = x_1 + x_2 = (i_{\text{ref}} - i_L) + (V_{\text{ref}} - \beta v_o) \\[2mm]
\quad\; = (K+1)\,[V_{\text{ref}} - \beta v_o] - i_L
\end{cases}
\tag{9.4}
$$

where $\bar{u} = 1 - u$ is the inverse logic of u; v_i denotes the instantaneous input voltage; i_C denotes the instantaneous capacitor current; and C and L denote the capacitance and inductance of the converters, respectively. Also, with respect to the converter, r_L and i_r are defined as the instantaneous load resistance and load current.

The equivalent control signal of the SM current controller when applied to the boost converter is obtained by solving $\frac{dS}{dt} = \alpha_1 \dot{x}_1 + \alpha_2 \dot{x}_2 + \alpha_3 \dot{x}_3 = 0$, which gives

$$
u_{\text{eq}} = 1 - \frac{K_2}{v_o} i_C - \frac{v_i}{v_o} + \frac{K_1}{v_o}\,[V_{\text{ref}} - \beta v_o] - \frac{K_3}{v_o} i_L
\tag{9.5}
$$

where

$$
K_1 = \frac{\alpha_3}{\alpha_1} L(K+1), \quad K_2 = \frac{\beta L}{C}\left(K + \frac{\alpha_2}{\alpha_1}\right), \quad \text{and } K_3 = \frac{\alpha_3}{\alpha_1} L
\tag{9.6}
$$

are the fixed gain parameters in the controller and u_{eq} is continuous and bounded by 0 and 1. Note that (9.5) can be expressed alternatively in terms of the original controlled state variables x_1 and x_2, i.e.,

$$
u_{\text{eq}} = 1 - \frac{K_2 \frac{v_o}{r_L} - v_i + K_3\,[V_{\text{ref}} - \beta v_o] - K_3\,(i_{\text{ref}} - i_L)}{K_2 i_L - v_o}
\tag{9.7}
$$

However, this will result in a complex form of equivalent control signal u_{eq}, which complicates the implementation of the controller. The use of instantaneous state variables i_C and i_L in the expression minimizes such complexity, as evidenced in the preceding section. It should also be mentioned that even though different forms of the equivalent control signal equation with respect to different choice of state variables can be obtained, they give essentially the same solution, i.e., $u_{\text{eq}} = f_1(x_1, x_2) = f_2(i_C, i_L)$.

9.3.4 Architecture of Controller

The SM current controller, which operates at a fixed frequency, is implemented through a pulse-width modulator. The controller is derived from the sliding surface (9.2) and then equating $u_{\text{eq}} = d$, where d is the duty ratio of a PWM

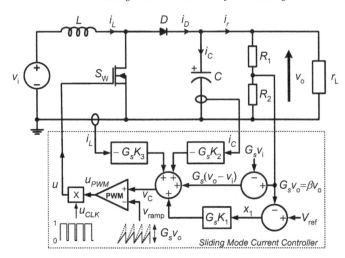

FIGURE 9.1
Sliding mode current controller for boost converters.

controller. The equations of the control law comprise a control signal v_c and a ramp signal $\hat{v}_{\text{ramp}}$, which are given by

$$\begin{cases} v_c = G_s K_1 \left[V_{\text{ref}} - \beta v_o\right] - G_s K_2 i_C - G_s K_3 i_L + G_s \left[v_o - v_i\right] \\ \hat{v}_{\text{ramp}} = G_s v_o \end{cases} \tag{9.8}$$

Here, a factor of $0 < G_s < 1$ has been intentionally introduced for down scaling the equation to a practical magnitude level that conforms to the chip-level's voltage standard of implementation. As in the PWM-based SM voltage controller for boost converters, a small protection circuitry is required to ensure that the duty ratio of the controller's output is always less than 1. This is achieved by multiplication of u_{PWM} of the pulse-width modulator with u_{CLK} of an impulse generator using a logic AND operator (see Fig. 8.14).

Figure 9.1 shows an overview of the SM current controller for the boost converter. The design of the controller is based on equation (9.8) and the assumption $\beta = G_s$. It should be noted that as in the fixed-frequency SM voltage controller described in Chapter 8, a ramp generator that varies its peak voltage with the change of output voltage is required for nonlinear compensation. However, unlike the SM voltage controller, the controller need not take into consideration the input voltage in its ramp voltage generation. This results in a simpler ramp generator circuitry in the latter than the former. It is also worth noting that two current sensors are required in the implementation of the controller. This is the main drawback of using this controller as compared to the SM voltage controller, which requires only the sensing of the capacitor current. Yet, the additional current sensing required on the inductor current is actually the main component constituting to a faster responding RHPZ converter system.

9.3.5 Existence Condition

So far, the hitting condition has been satisfied by choosing the switching function appropriately. As for the existence condition, it can be obtained by inspecting the local reachability condition $\lim_{S \to 0} S \cdot \frac{dS}{dt} < 0$, which with the substitutions of (9.2) and its time derivative, gives

$$
\begin{cases}
\alpha_1 \left[-\frac{\beta K}{C} i_C - \frac{v_i}{L} \right] - \alpha_2 \frac{\beta}{C} i_C + \alpha_3 \left((K+1) \left[V_{\text{ref}} - \beta v_o \right] - i_L \right) < 0 \\
\alpha_1 \left[-\frac{\beta K}{C} i_C - \frac{v_i - v_o}{L} \right] - \alpha_2 \frac{\beta}{C} i_C + \alpha_3 \left((K+1) \left[V_{\text{ref}} - \beta v_o \right] - i_L \right) > 0
\end{cases}
\quad . (9.9)
$$

Assuming the controller is designed with a static sliding surface to meet the existence conditions for steady-state operations (equilibrium point) [26, 49, 94], and with the consideration of equation (9.6), then (9.9) can be simplified as

$$
\begin{cases}
0 < v_{i(\text{min})} - K_1 \left[V_{\text{ref}} - \beta v_{o(\text{SS})} \right] + K_2 i_{C(\text{min})} + K_3 i_{L(\text{max})} \\
v_{i(\text{max})} - K_1 \left[V_{\text{ref}} - \beta v_{o(\text{SS})} \right] + K_2 i_{C(\text{max})} + K_3 i_{L(\text{min})} < v_{o(\text{SS})}
\end{cases}
\quad (9.10)
$$

where $v_{i(\text{max})}$ and $v_{i(\text{min})}$ denotes the maximum and minimum input voltages respectively; $v_{o(\text{SS})}$ denotes the expected steady-state output voltage which is basically a DC parameter of a small error from the desired reference voltage V_{ref}; and $i_{L(\text{max})}$, $i_{L(\text{min})}$, $i_{C(\text{max})}$, and $i_{C(\text{min})}$ are respectively the maximum and minimum inductor and capacitor currents when the converter is operating at full-load condition. Figure 9.2 illustrates the physical representation of these parameters.

Finally, the selection of the controller's gain parameters K_1, K_2, and K_3 must comply with the set of inequalities in (9.10). This assures the existence of the SM operation at least in the small region of the origin for all operating conditions within the designated input and load range.

9.3.6 Stability Condition

Unlike in SM voltage controllers where the selection of the sliding coefficients (control gains) to satisfy the stability condition is automatically performed by designing the system for some desired dynamic properties [92, 94], the same approach cannot be adopted for designing the SM current controller. This is because the motion equation (derived from $S = 0$) of the controller, which composes both the current and voltage state variables, is highly nonlinear and cannot be easily solved analytically. A different approach based on the *equivalent control method* is adopted. This involves first deriving the *ideal sliding dynamics* of the system, then performing an analysis on the *equilibrium point*, and finally deriving the condition for maintaining stability [49, 50]. Refer to Chapter 1 for a summary of the derivation procedure.

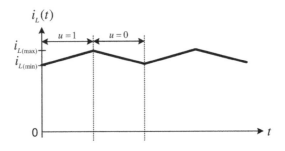

(a) Inductor current waveform

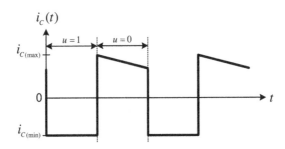

(b) Capacitor current waveform

FIGURE 9.2
Current waveforms of boost converters.

9.3.6.1   Ideal Sliding Dynamics

The replacement of $\bar{u}$ by $\bar{u}_{\text{eq}}$ (so-called equivalent control method) in the original boost converter's description under continuous conduction mode operation converts the discontinuous system into an ideal SM continuous system:

$$\begin{cases} \dfrac{di_L}{dt} = \dfrac{v_{\text{i}}}{L} - \dfrac{v_{\text{o}}}{L}\bar{u}_{\text{eq}} \\[2ex] \dfrac{dv_{\text{o}}}{dt} = \dfrac{i_L}{C}\bar{u}_{\text{eq}} - \dfrac{v_{\text{o}}}{r_{\text{L}}C} \end{cases}. \tag{9.11}$$

Then, the substitution of (9.7) into (9.11) gives

$$\begin{cases} \dfrac{di_L}{dt} = \dfrac{v_{\text{i}}}{L} - \dfrac{v_{\text{o}}}{L}\dfrac{K_2\frac{v_{\text{o}}}{r_{\text{L}}} - v_{\text{i}} + K_3\left[V_{\text{ref}} - \beta v_{\text{o}}\right] - K_3\left(i_{\text{ref}} - i_L\right)}{K_2 i_L - v_{\text{o}}} \\[3ex] \dfrac{dv_{\text{o}}}{dt} = \dfrac{i_L}{C}\dfrac{K_2\frac{v_{\text{o}}}{r_{\text{L}}} - v_{\text{i}} + K_3\left[V_{\text{ref}} - \beta v_{\text{o}}\right] - K_3\left(i_{\text{ref}} - i_L\right)}{K_2 i_L - v_{\text{o}}} - \dfrac{v_{\text{o}}}{r_{\text{L}}C} \end{cases} \tag{9.12}$$

which represents the ideal sliding dynamics of the SM current controlled boost converter.

9.3.6.2 Equilibrium Point Analysis

Assume that there exists a stable equilibrium point on the sliding surface on which the ideal sliding dynamics eventually settles. At equilibrium, we have $\frac{di_L}{dt} = \frac{dv_o}{dt} = 0$. Thus, the steady-state equation is

$$I_L = \frac{V_o^2}{V_i R_L} \tag{9.13}$$

where I_L, V_o, V_i, and R_L represents the inductor current, output voltage, input voltage, and load resistance at steady-state equilibrium, respectively.

9.3.6.3 Linearization of Ideal Sliding Dynamics

Next, linearization of the ideal sliding dynamics around the equilibrium point transforms equation (9.12) into

$$\begin{cases} \dfrac{d\tilde{i}_L}{dt} = a_{11}\tilde{i}_L + a_{12}\tilde{v}_o \\ \dfrac{d\tilde{v}_o}{dt} = a_{21}\tilde{i}_L + a_{22}\tilde{v}_o \end{cases} \tag{9.14}$$

where

$$\begin{cases} a_{11} = \dfrac{K_3 V_i R_L}{K_2 L V_o - L V_i R_L} \\[2mm] a_{12} = \dfrac{K_1 \beta V_i R_L - 2K_2 V_i + \frac{V_i^2 R_L}{V_o}}{K_2 L V_o - L V_i R_L} \\[2mm] a_{21} = \dfrac{K_2 V_i - \frac{V_i^2 R_L}{V_o} - K_3 V_o}{K_2 V_o C - C V_i R_L} \\[2mm] a_{22} = \dfrac{\frac{K_2 V_o}{R_L} - K_1 \beta V_o}{K_2 V_o C - C V_i R_L} - \dfrac{1}{R_L C} \end{cases} \tag{9.15}$$

The derivation has been performed by assuming that the system operates at steady state, i.e., $v_i = V_i$, $r_L = R_L$, $V_{ref} - \beta V_o = 0$, and $I_{ref} - I_L = 0$, and that $I_L \gg \tilde{i}_L$ and $V_o \gg \tilde{v}_o$. The characteristic equation of this linearized system can be written as

$$s^2 - (a_{11} + a_{22})s + a_{11}a_{22} - a_{12}a_{21} = 0. \tag{9.16}$$

The system will be stable if the following conditions are satisfied

$$a_{11} + a_{22} < 0$$
$$a_{11}a_{22} - a_{12}a_{21} > 0. \tag{9.17}$$

For the case of $a_{11} + a_{22} < 0$, the condition for stability is

$$\frac{K_3 C V_i R_L - K_1 L \beta V_o + L V_i}{K_2 V_o - V_i R_L} < 0, \tag{9.18}$$

i.e.,

$$\begin{cases} K_3 \dfrac{V_i R_L C}{L \beta V_o} + \dfrac{V_i}{\beta V_o} < K_1 & \text{when } K_2 > \dfrac{V_i R_L}{V_o} \\[3mm] K_3 \dfrac{V_i R_L C}{L \beta V_o} + \dfrac{V_i}{\beta V_o} > K_1 & \text{when } K_2 < \dfrac{V_i R_L}{V_o} \end{cases} \tag{9.19}$$

As for the case of $a_{11} a_{22} - a_{12} a_{21} > 0$, the condition for stability is

$$2K_3 V_o^3 (K_2 - K_1 \beta R_L) + V_i V_o^2 K_2 (K_1 \beta R_L - 2K_2) \tag{9.20}$$
$$+ V_i^2 V_o R_L (3K_2 - K_1 \beta R_L) - V_i^3 R_L^2 > 0$$

In essence, the existence condition (9.10) and the stability conditions (9.19) and (9.20) form the basis for the selection and design of the control gains of the SM current controller in terms of the converter's specifications. Satisfaction of these conditions assures the *closed-loop stability* of the system.

9.3.7 An Empirical Approach of Selecting the Sliding Coefficients

To alleviate the difficulty in designing the control gains that satisfy the various conditions, computer simulation and experiments have been performed to study the effects of the various control gains on the response of the output voltage. It is observed that

a) an increment in K_1 improves the steady-state regulation but causes the transient response to be more oscillatory with a higher overshoot, thus prolonging the steady-state settling time (see Fig. 9.3(a));

b) an increment in K_2 improves the steady-state regulation and also makes the transient response less oscillatory with a lower overshoot, thus shortening the steady-state settling time (see Fig. 9.3(b)). However, the range of adjustable values for K_2 is very small (limited by the bidirectional capacitor current);

c) a relatively small increment in K_3 can have a moderate reduction in the oscillation of the transient response and also significant shortening of the steady-state settling time. However, the steady-state regulation is deteriorated (see Fig. 9.3(c)).

Hence, in accordance with these observations, a heuristic but practical approach of designing the control gains is to first select the highest possible values of K_1 and K_2, and an arbitrary low value of K_3, without compromising

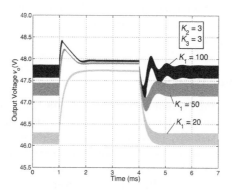

(a) Different K_1 settings

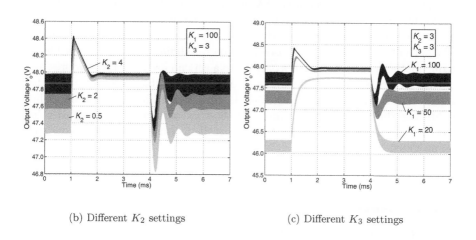

(b) Different K_2 settings   (c) Different K_3 settings

FIGURE 9.3
Simulated waveforms of the output voltage v_o of a boost converter under the
SM current controller at various control gains settings.

the existence or stability condition for the full-load condition. Then, having
the converter operated at the nominal input voltage and applying step load
changes between the minimum and maximum loading conditions, the values
of K_1 and K_3 can be adjusted to fit the converter's performance to the desired
profile. As in conventional controllers, such tuning of the control parameters is
typically required in the initial prototyping stage to ensure that the converter
responses in the desired manner. However, an advantage of the SM current
controller over a conventional controller is that once the parameters are de-
cided for the nominal condition, the large-signal property of the SM current
controller will ensure a consistent dynamical behavior for all operating condi-

tions, as in the SM voltage controller. This eliminates the need for re-tuning the parameters to suit other operating conditions, as would be required in conventional controllers.

9.3.8 Additional Remarks

The fixed-frequency SM current controller inherits the other advantages of controlling current, namely, the over-current and audio-susceptibility protection properties. Additionally, being a form of nonlinear controller, it is capable of handling large-signal perturbations with excellent consistency in its dynamical responses. In other words, the main application of this controller is to ensure fast response in RHPZ converters over a wide range of operating conditions. This is usually not achievable by conventional current-mode controllers. Although additional current sensor and a slightly more complicated circuit architecture are required, SM current controllers can be the preferred choice for application demanding very high dynamical performance.

On the other hand, it is clearly seen from the simulation results that the SM current controller, which uses an integral sliding surface, is capable of reducing the steady-state errors only up to a certain level. Further suppression of these errors is not possible unless an additional double integral term of the state variables, i.e., $\int (\int [x_1 + x_2] dt) dt$, is incorporated in the controller. This is because the increased order (pole) of the controller generally improves the steady-state accuracy of the system. A detailed discussion on this issue will be provided in Chapter 11.

9.4 Results and Discussions

In this section, simulation[1] and experimental results of the SM current controller are provided to validate the theoretical design. Both the simulation program and the hardware prototype are developed from equation (9.8) for a 100 W boost converter with specifications shown in Table 9.1. The control parameters adopted are $V_{\text{ref}} = 6$ V, $G_s = \beta = \frac{1}{8}$ V/V, $K_1 = 80$ V/V, $K_2 = 3.12$ V/A, and $K_3 = 2.67$ V/A. They are chosen to comply with the design restrictions in (9.10), (9.19), and (9.20), and have been fine tuned to respond to a desired regulation and dynamic response. Figure 9.4 shows the full schematic diagram of the experimental prototype, which is constructed using simple analog ICs and operational amplifiers (also see Fig. 9.1).

---

[1]The simulation is performed using Matlab/Simulink. The step size taken for all simulations is 10 ns.

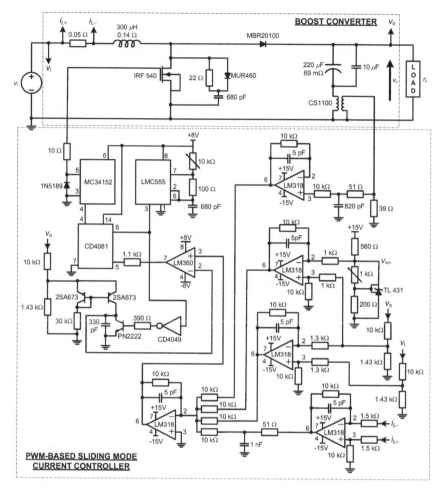

FIGURE 9.4
Full schematic diagram of the SM current controlled boost converter prototype.

9.4.1 Regulation Performance

Figure 9.5 shows the steady-state waveforms of the SM current controlled boost converter operating at the nominal input voltage. The control signal v_c is derived from analog computation of the instantaneous feedback signals v_o, v_i, i_L, and i_C using the expression described in (9.8). Also, the peak amplitude of the ramp signal v_{ramp} is generated to follow $G_s v_o$ as in (9.8). Both v_c and v_{ramp} are fed into a pulse-width modulator to generate the gate pulse u for controlling the switching action of the boost converter. Except for some ringing noise and a small time delay in the experimentally captured v_c waveforms,

TABLE 9.1

Specifications of boost converter.

Description	Parameter	Nominal Value
Input voltage	v_i	24 V
Capacitance	C	230 μF
Capacitor ESR	r_c	69 mΩ
Inductance	L	300 μH
Inductor resistance	r_l	0.14 Ω
Switching frequency	f_S	200 kHz
Minimum load resistance (full load)	$r_{L(min)}$	24 Ω
Maximum load resistance (10% load)	$r_{L(max)}$	240 Ω
Desired output voltage	V_{od}	48 V

TABLE 9.2

Load regulation property: output voltage at nominal operating condition $v_i = 24$ V and $r_L = 24\ \Omega$ is $v_{o(nominal)} = 47.45$ V.

Input voltage	$\Delta v_o = v_{o(240\ \Omega)} - v_{o(24\ \Omega)}$	$\frac{\Delta v_o}{v_{o(nominal)}}$ x 100%
$v_i = 20$ V	1.13 V	2.38% of $v_{o(nominal)}$
$v_i = 24$ V	0.82 V	1.73% of $v_{o(nominal)}$
$v_i = 28$ V	0.35 V	0.74% of $v_{o(nominal)}$

TABLE 9.3

Line regulation property: output voltage at nominal operating condition $v_i = 24$ V and $r_L = 24\ \Omega$ is $v_{o(nominal)} = 47.45$ V.

Load condition	$\Delta v_o = v_{o(v_i=20\ V)} - v_{o(v_i=28\ V)}$	$\frac{\Delta v_o}{v_{o(nominal)}}$ x 100%
Minimum load (240 Ω)	0.40 V	0.84% of $v_{o(nominal)}$
Half load (48 Ω)	0.27 V	0.57% of $v_{o(nominal)}$
Full load (24 Ω)	0.14 V	0.29% of $v_{o(nominal)}$

the simulation and experimental results are in good agreement with the theoretical prediction. The slight discrepancy is mainly due to the non-ideality of practical sensors and analog components, which are not modeled in the simulation program. Yet, it is worth mentioning that as in average current-mode control, the ringing noise has much less influence on the generated control pulse compared to the case of the peak current-mode control.

Tabulations of the data in terms of the load and line regulation properties are also given in Tables 9.2 and 9.3 respectively. According to Table 9.2, the maximum load-regulation error occurs at $v_i = 20$ V, with a deviation of 2.38% from $v_{o(nominal)}$. Similarly, it can be found from Table 9.3 that the maximum line-regulation error occurs at minimum load $r_L = 240\ \Omega$, with a deviation of 0.84% from $v_{o(nominal)}$.

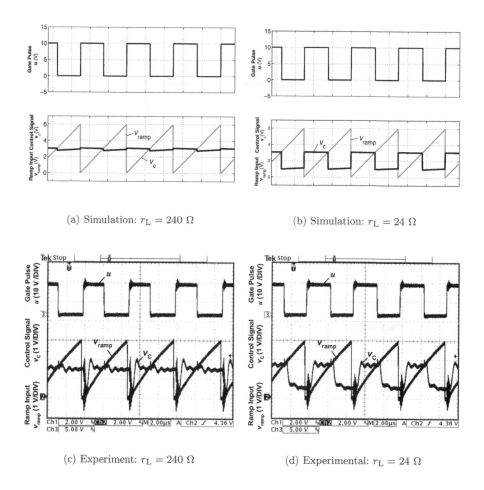

(a) Simulation: $r_L = 240\ \Omega$ (b) Simulation: $r_L = 24\ \Omega$

(c) Experiment: $r_L = 240\ \Omega$ (d) Experimental: $r_L = 24\ \Omega$

FIGURE 9.5
Steady-state waveforms of control signal v_c, input ramp v_{ramp}, and generated gate pulse u of the SM current controlled boost converter operating at nominal input voltage $v_i = 24$ V.

9.4.2 Dynamic Performance

The capability of the controller in handling large-signal disturbances is compared to that of a UC3843 peak current-mode controller. Comparison with the average current-mode controller is redundant in this case since it offers similar dynamic characteristic as that of the peak current-mode controller under large-signal disturbances. Moreover, average current-mode controller is not often adopted in practice.

Figures 9.6(a)–9.6(c) show the experimental waveforms of the peak current-mode controlled boost converter operating with the load resistance

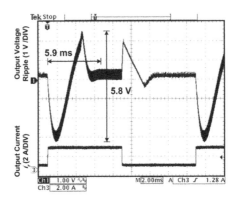

(a) $v_i = 20$ V (0.2/2.0 A)

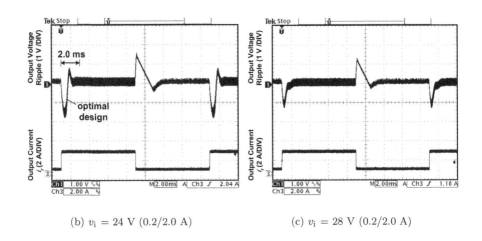

(b) $v_i = 24$ V (0.2/2.0 A)   (c) $v_i = 28$ V (0.2/2.0 A)

FIGURE 9.6
Experimental waveforms of output voltage ripple $\widetilde{v}_o$ and output current i_r of the boost converter with the peak current-mode controller operating at input voltage 20 V (minimum), 24 V (nominal), and 28 V (maximum), and alternating between load resistances 24 Ω (minimum) and 240 Ω (maximum).

stepping between $r_L = 24$ Ω and $r_L = 240$ Ω for various input voltages. The converter has been optimally tuned for the condition shown in Fig. 9.6(b). It can be observed that the system exhibits different transient settling times for different operating conditions. In the worst-case operating condition: $v_i = 20$ V and the output current alternates between 0.2 A and 2.0 A, the system has a settling time of 5.9 ms and a relatively high voltage ripple of 5.8 V (see Fig. 9.6(a)), which deviate greatly from the optimal performance of 2 ms and a voltage ripple of 2.6 V (see Fig. 9.6(b)). This large variation in the dy-

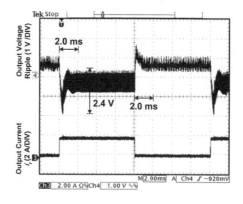

(a) $v_i = 20$ V (0.2/2.0 A)

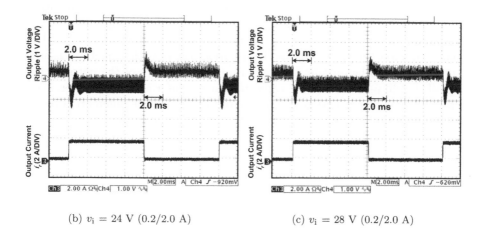

(b) $v_i = 24$ V (0.2/2.0 A) (c) $v_i = 28$ V (0.2/2.0 A)

FIGURE 9.7
Experimental waveforms of output voltage ripple $\tilde{v}_o$ and output current i_r of the boost converter with the SM current-mode controller operating at input voltage 20 V (minimum), 24 V (nominal), and 28 V (maximum), and alternating between load resistances 24 Ω (minimum) and 240 Ω (maximum).

namic performance at various operating conditions is expected from the peak current-mode controller, which is semi-nonlinear and its compensation network has been designed on the basis of a linearized small-signal model, and hence is only optimal for a specific operating condition.

Figures 9.7(a)–9.7(c) show the dynamic behavior of the SM current controlled boost converter at various input and load conditions. It can be seen that the transient settling time is around 2.0 ms for all operating conditions, including the worst-case operating condition which has a voltage ripple of around 2.4 V (see Fig. 9.7(a)). This is close to the optimally designed point, which has a voltage ripple of 1.8 V (see Fig. 9.7(b)). Comparing to the peak current-mode controller, which in its worst case has a settling time of 5.9 ms and a voltage ripple of 5.8 V, the advantage of the SM current-mode controller in giving a fast and consistent dynamical response over a wide range of operating conditions is clearly demonstrated.

On the other hand, the output voltage ripple of the boost converter is higher with the SM current controller than the peak current-mode controller. The higher voltage ripple is caused by the impedance of the additional current sensor CS 1100 located at the output capacitor filter path in the scheme. Typically, it can be reduced using a current sensor of lower impedance. Otherwise, if required, the indirect sensing of the capacitor current, which totally removes the requirement for the capacitor current sensor, will restore the voltage ripple back to its original state. There are two methods of doing this. One method is by sensing the inductor current, and then extracting the ac component of the turn-off (i.e., $1 - D$) period of the inductor current signal. This operation will produce an information that is identical to that of the capacitor current signal. The disadvantage of using this method is the requirement of an additional circuitry for processing the information. On the other hand, capacitor current can also be obtained by performing a derivative of the output voltage. This will also provide the capacitor current information. However, such operation requires careful noise filtering of the signal since differentiators are highly noise sensitive and may distort the required information.

10

Sliding Mode Control with a Reduced-State Sliding Manifold for High-Order Converters

CONTENTS

10.1 Introduction

In this chapter, we will continue our discussion from the previous chapter on the SM control design for power converters with a nonlinear sliding manifold. However, here, the discussion will be focused on how a reduced-state sliding manifold may be used for the control of a high-order converter like the Ćuk converter. Unlike a full-state SM controller, which exercises full control over

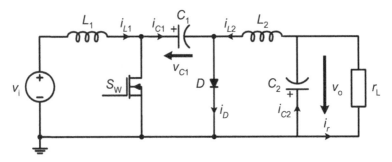

FIGURE 10.1
A Ćuk converter.

how all the independent state variables of a system behave, a reduced-state controller controls only some available state variables such that while the system can still achieve important characteristics like stability and reasonable transient response, there is no control over the precise response of all the individual state variables. On the other hand, the major advantage of using a reduced-state sliding manifold over a full-state sliding manifold is the simplicity of the design and implementation of the SM controller, especially for power converters of high order. As an example, the discussion will be focused on a type of pulse-width modulation (PWM)-based reduced-state SM controller for the Ćuk converter.

10.2 Review of Conventional Sliding Mode Controllers for Ćuk Converters

10.2.1 State-Space Model of Ćuk Converter

Figure 10.1 shows the Ćuk converter [18], which consists of an input inductor L_1, power switch S_W, energy transfer capacitor C_1, free-wheeling diode D, output inductor L_2, and filter capacitor C_2. The operating principle of the Ćuk converter has been widely discussed and will be omitted here. Briefly, the function of this converter is to transfer electrical energy from the input voltage supply v_i to the output load r_L at a voltage level that can be higher or lower than the input supply through the energy transfer capacitor. The state variables are input inductor current i_{L1}, energy transfer capacitor voltage v_{C1}, output inductor current i_{L2}, and output voltage v_o. Under continuous conduction mode of operation, the state-space model of the Ćuk converter can be expressed as

$$
\begin{cases}
L_1 \dfrac{di_{L1}}{dt} = v_\mathrm{i} - \bar{u} v_{C1} \\[2mm]
L_2 \dfrac{di_{L2}}{dt} = u v_{C1} - v_\mathrm{o} \\[2mm]
C_1 \dfrac{dv_{C1}}{dt} = \bar{u} i_{L1} - u i_{L2} \\[2mm]
C_2 \dfrac{dv_\mathrm{o}}{dt} = i_{L2} - \dfrac{v_\mathrm{o}}{r_\mathrm{L}}
\end{cases}
\tag{10.1}
$$

where $u = (0, 1)$ is the logic state of the power switch S_W, and $\bar{u} = 1 - u$ is the inverse logic of u. Optimal control of the power conversion can be achieved through the appropriate control of the state variables.

10.2.2 Full-State SM Controller

A full-state SM controller controls all the independent state variables in the system. Since the Ćuk converter is a fourth-order network (four independent storage elements), its full-state SM controller involves the control of all the four state variables. A description of this full-state SM controller is illustrated in Malesani *et al.* [46]. The controller requires the sensing of the two voltage variables, i.e., v_{C1} and v_o, and the sensing of the two current variables, i.e., i_{L1} and i_{L2}. Additionally, it requires the generation of four signals for referencing the state variables, i.e., $v_{C1(\mathrm{ref})}$, $v_{\mathrm{o(ref)}}$, $i_{L1(\mathrm{ref})}$, and $i_{L2(\mathrm{ref})}$. Furthermore, the switching frequency of this controller is not fixed and is dependent on the system's input and output conditions. Thus, various reduced-state SM controllers have been proposed to reduce complexity.

10.2.3 Reduced-State SM Controller

The purpose of reducing the number of state variables in the controller is to minimize the number of state variables that need to be sensed and the number of signals that have to be generated in the controller. Previously proposed reduced-state SM controllers can be categorized into two types. One type uses only the output voltage as the controlled variable, i.e., voltage-mode control SM controller, and the other uses both the output voltage and the inductor current as the controlled variables, i.e., current-mode control SM controller.

10.2.3.1 Voltage-Mode Control SM Controller

Two types of reduced-state SM voltage controllers have been proposed for the Ćuk converter by Mahdavi *et al.* [44] (simple form) and [45] (with neural networks). The main advantages of these controllers are that they require the sensing of only two voltage variables, i.e., v_i and v_o; one reference signal, i.e., the output voltage reference $v_{\mathrm{o(ref)}}$; and they operate at constant frequency. However, the drawbacks are that they require numerous complex mathematical operations, which may reduce the practicality of the proposed controllers. Furthermore, it is known that the application of voltage-mode control (linear or nonlinear types) to non-minimum phase converters will result in sub-optimal performance with rather narrow bandwidth, i.e., slow response. It is, therefore, necessary to take into consideration the current-mode control variables when formulating the SM controllers for non-minimum phase converters.

10.2.3.2   Current-Mode Control SM Controller

An original form of the reduced-state SM current controllers is illustrated in Malesani *et al.* [46]. It is shown that a second-order SM controller with only two state variables (input inductor current and output voltage) is sufficient to ensure the stability of the Ćuk converter. The advantages of this controller are that it requires the sensing of only the output voltage v_o and the input inductor current i_{L1}; it requires only the output voltage reference $v_{o(\text{ref})}$; and that it gives optimal performance similar to that of the full-state SM controllers. However, despite its excellent features, it has a major drawback of being a variable-frequency controller. Hence, it is still necessary to operate reduced-state SM current controllers at constant frequency before they can be put to practical use.

10.3   Constant-Frequency Reduced-State Sliding Mode Current Controller

The constant-frequency reduced-state SM current controller employs both the output voltage error and the inductor current error as the controlled state variables. This is similar to the reduced-state SM controller reported by Malesani *et al.* [46]. However, the reference inductor current profile of the controller is generated using the amplified output voltage error, i.e.,

$$i_{\text{ref}} = K \left[V_{\text{ref}} - \beta v_o \right] \tag{10.2}$$

where V_{ref} and v_o denote the reference and instantaneous output voltages respectively; β denotes the feedback network ratio; and K is the amplified gain of the voltage error. This is similar to that of the conventional PWM current-mode controllers. It should be mentioned that the use of integral control here is optional. A large value of K in the outer voltage loop is sufficient to ensure an excellent regulation of the output voltage with a negligible steady-state error.

10.3.1 Sliding Surface

We denote S as the sliding surface, which can be chosen as a linear combination of the four state variables, i.e.,

$$S = \alpha_1 x_1 + \alpha_2 x_2 + \alpha_3 x_3 + \alpha_4 x_4 \qquad (10.3)$$

where α_1, α_2, α_3, and α_4 represents the sliding coefficients. The switching function u, which represents the logic state of power switch S_W, can be defined as $u = \frac{1}{2}(1 + \text{sign}(S))$. Here, the adopted controlled state variables are the *current error* x_1, the *voltage error* x_2, the *integral of the current and the voltage errors* x_3, and the *double integral of the current and the voltage errors* x_4, which are expressed as

$$\begin{cases} x_1 = i_{\text{ref}} - i_{L1} \\ x_2 = V_{\text{ref}} - \beta v_{\text{o}} \\ x_3 = \int [x_1 + x_2]\, dt \\ x_4 = \int (\int [x_1 + x_2]\, dt) dt \end{cases} \qquad (10.4)$$

Ideally, at an infinitely high switching frequency, only the controlled variables x_1 and x_2 are required in the SM current controller to ensure that both the output voltage and inductor current follow exactly their references, i.e., $v_{\text{o}} = V_{\text{ref}}$ and $i_{L1} = i_{\text{ref}}$. However, in the case of finite-frequency or fixed-frequency SM controllers, the control is imperfect. Significant amount of steady-state errors exist in both the output voltage error and the inductor current error. The integral term x_3 and the double integral term x_4 are adopted to suppress the steady-state errors of the system due to computational errors in the controller as inherited from implementation in the indirect (PWM) form [92]. It is found that the adoption of the integral term x_3 alone *can reduce, but not fully eliminate*, the steady-state error of the converters. The idea of having a double integral term x_4 of the original state variables to correct the steady-state errors is based on the control principle that the increased order of the controller improves the steady-state accuracy of the system. A detailed discussion on this issue will be provided in Chapter 11.

10.3.2 Dynamical Model of Controller/Converter System and Its Equivalent Control

The substitution of the Ćuk converter's behavioral model under continuous conduction mode into the time differentiation of (10.4) gives the dynamical model of the system as

$$
\begin{cases}
\dot{x}_1 = \dfrac{d[i_{\text{ref}} - di_{L1}]}{dt} = -\dfrac{K\beta}{C_2}i_{C2} - \dfrac{v_{\text{i}} - \bar{u}v_{C1}}{L_1} \\[2mm]
\dot{x}_2 = \dfrac{d[V_{\text{ref}} - \beta v_{\text{o}}]}{dt} = -\dfrac{\beta}{C_2}i_{C2} \\[2mm]
\dot{x}_3 = x_1 + x_2 = (K+1)\left[V_{\text{ref}} - \beta v_{\text{o}}\right] - i_{L1} \\[2mm]
\dot{x}_4 = \int (x_1 + x_2)\, dt = \int [(K+1)\left[V_{\text{ref}} - \beta v_{\text{o}}\right] - i_{L1}]dt
\end{cases}
\qquad (10.5)
$$

The equivalent control signal of the SM current controller when applied to the Ćuk converter is obtained by solving $\frac{dS}{dt} = \alpha_1 \dot{x}_1 + \alpha_2 \dot{x}_2 + \alpha_3 \dot{x}_3 + \alpha_4 \dot{x}_4 = 0$, which gives

$$
u_{\text{eq}} = 1 - \frac{1}{v_{C1}} \left[\frac{\beta L_1}{C_2} \left(K + \frac{\alpha_2}{\alpha_1} \right) i_{C2} + v_{\text{i}} - \frac{\alpha_3}{\alpha_1} L_1 (x_1 + x_2) \right.
$$

$$
\left. - \frac{\alpha_4}{\alpha_1} L_1 \int (x_1 + x_2)\, dt \right], \qquad (10.6)
$$

where u_{eq} is continuous and bounded by 0 and 1. Note that (10.5) can be expressed in terms of the original controlled state variables x_1 and x_2. However, this will result in a complex form of equivalent control signal u_{eq}, which complicates the implementation of the controller. The use of instantaneous state variables i_{C2} and i_{L1} in the expression simplifies the implementation, as explained in the preceding section. It should also be mentioned that even though different forms of the equivalent control signal equation with respect to different choices of state variables can be obtained, they fundamentally gives the same solution, i.e., $u_{\text{eq}} = f_1(x_1, x_2) = f_2(i_{C2}, i_{L1})$.

10.3.3 Architecture of Controller

The SM current controller, which operates at a fixed frequency, is implemented through a pulse-width modulator. The control equation can be expressed as

$$
\begin{cases}
v_{\text{c}} = G_s (v_{C1} - v_{\text{i}}) - K_3 i_{C2} + K_1 (V_{\text{ref}} - \beta v_{\text{o}}) + K_2 \int (V_{\text{ref}} - \beta v_{\text{o}})\, dt \\[2mm]
\quad + K_1 (K [V_{\text{ref}} - \beta v_{\text{o}}] - i_{L1}) + K_2 \int (K [V_{\text{ref}} - \beta v_{\text{o}}] - i_{L1})\, dt \\[2mm]
\hat{v}_{\text{ramp}} = G_s v_{C1}
\end{cases}
\qquad (10.7)
$$

where

$$
K_1 = G_s \frac{\alpha_3}{\alpha_1} L_1, \quad K_2 = G_s \frac{\alpha_4}{\alpha_1} L_1, \quad \text{and} \quad K_3 = G_s \frac{\beta L_1}{C_2} \left(K + \frac{\alpha_2}{\alpha_1} \right) \qquad (10.8)
$$

are the fixed gain parameters of the controller. A factor of $0 < G_s < 1$ has been introduced to scale down the equation to conform to the chip level voltage standard. With this, the equivalent control signal in (10.6) can be expressed as

$$
u_{\text{eq}} = 1 - \frac{K_3 i_{C2} + G_s v_{\text{i}} - K_1 (x_1 + x_2) - K_2 \int (x_1 + x_2)\, dt}{G_s v_{C1}}. \qquad (10.9)
$$

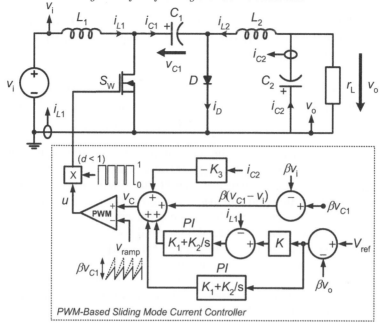

FIGURE 10.2

Reduced-state PWM-based SM current controller for Ćuk converters.

Figure 10.2 shows the analog implementation of the reduced-state PWM-based SM current controller for the Ćuk converters. The design of the controller is based on equations (10.7) and (10.8), and the assumption $\beta = G_s$. Note that a square wave signal and a multiplier have been included at the output of the PWM block. This is to ensure that the duty ratio of the controller's output is always less than 1. This is important since the boost and buck-boost types of converters cannot operate with a switching signal u that has a duty ratio equals to 1. Finally, it is worth mentioning that the controller requires two current sensors and three voltage sensors for its implementation. This added complexity is the necessary tradeoff for transforming the controller from variable frequency to fixed frequency. Hence, the choice among the various reduced-state SM controllers is a decision between converters' specifications and performance requirements.

10.3.4 Existence Condition

The existence condition can be obtained from $\lim_{S \to 0} S \cdot \frac{dS}{dt} < 0$, which with the substitutions of (10.3) and its time derivative, gives

$$\begin{cases} -K_3 i_{C2} + K_1 (x_1 + x_2) + K_2 x_3 - \beta v_i < 0 \\ -K_3 i_{C2} + K_1 (x_1 + x_2) + K_2 x_3 + \beta (v_{C1} - v_i) > 0 \end{cases} . \tag{10.10}$$

Assuming the controller is designed with a static sliding surface to meet the existence conditions for steady-state operations (equilibrium point) [26, 49], then (10.10) can be simplified as

$$
\begin{cases}
-K_3 i_{C2(\min)} + K_1 \left(x_{1(\max)} + x_{2(\max)} \right) + K_2 x_{3(\max)} < \beta v_{i(\min)} \\
K_3 i_{C2(\max)} - K_1 \left(x_{1(\min)} + x_{2(\min)} \right) - K_2 x_{3(\min)} \\
\qquad\qquad\qquad\qquad < \beta \left(v_{i(\max)} - v_{C1(ss)} \right)
\end{cases}
\tag{10.11}
$$

where $v_{i(\max)}$ and $v_{i(\min)}$ denote the maximum and minimum input voltages respectively; $v_{C1(ss)}$ is the output voltage of energy transfer capacitor at steady state; $i_{C2(\max)}$, and $i_{C2(\min)}$ are respectively the maximum and minimum filter capacitor currents at full-load condition; $x_{1(\max)}$ and $x_{1(\min)}$ are respectively the maximum and minimum steady-state current errors; $x_{2(\max)}$ and $x_{2(\min)}$ are respectively the maximum and minimum steady-state voltage errors, which in this case are basically the inverse functions of the output voltage ripples; and $x_{3(\max)}$ and $x_{3(\min)}$ are respectively the maximum and minimum integrals of the combination of steady-state voltage and current errors. These parameters can all be calculated from the design specifications of the converter. Alternatively, computer simulation of the converter under an ideal open-loop control which gives negligible output voltage steady-state error can be performed to obtain these values. Thus, the compliance of the inequalities in (10.11) assures the existence of the SM operation at least in the small region of the origin for all operating conditions within the designated input and load range.

10.3.5 Stability Condition

For the controller, the stability condition can be derived by first obtaining the *ideal sliding dynamics* of the system, and then performing an analysis on its *equilibrium point* [49, 50].

10.3.5.1 Ideal Sliding Dynamics

The replacement of u by u_{eq} and $\bar{u}$ by $\bar{u}_{\mathrm{eq}}$ (equivalent control method) in the original Ćuk converter's description under continuous conduction mode operation converts the discontinuous system into an ideal SM continuous system:

$$
\begin{cases}
\dfrac{di_{L1}}{dt} = \dfrac{v_i}{L_1} - \dfrac{v_{C1}}{L_1} \bar{u}_{\mathrm{eq}} \\[2mm]
\dfrac{dv_{C1}}{dt} = \dfrac{i_{L1}}{C_1} \bar{u}_{\mathrm{eq}} - \dfrac{i_{L2}}{C_1} u_{\mathrm{eq}} \\[2mm]
\dfrac{di_{L2}}{dt} = \dfrac{v_{C1}}{L_2} u_{\mathrm{eq}} - \dfrac{v_o}{L_2} \\[2mm]
\dfrac{dv_o}{dt} = \dfrac{i_{L2}}{C_2} - \dfrac{v_o}{r_L C_2}
\end{cases}
\tag{10.12}
$$

Then, the substitution of the equivalent control signal (10.9) into (10.12) gives

$$
\begin{cases}
\frac{di_{L1}}{dt} = \frac{v_i}{L_1} - \frac{v_{C1}}{L_1}\left[\frac{K_3 i_{C2}+G_s v_i - K_1(x_1+x_2)-K_2\int(x_1+x_2)dt}{G_s v_{C1}}\right] \\
\frac{dv_{C1}}{dt} = \frac{i_{L1}}{C_1}\left[\frac{K_3 i_{C2}+G_s v_i - K_1(x_1+x_2)-K_2\int(x_1+x_2)dt}{G_s v_{C1}}\right] \\
\qquad - \frac{i_{L2}}{C_1}\left[1 - \frac{K_3 i_{C2}+G_s v_i - K_1(x_1+x_2)-K_2\int(x_1+x_2)dt}{G_s v_{C1}}\right] \\
\frac{di_{L2}}{dt} = \frac{v_{C1}}{L_2}\left[1 - \frac{K_3 i_{C2}+G_s v_i - K_1(x_1+x_2)-K_2\int(x_1+x_2)dt}{G_s v_{C1}}\right] - \frac{v_o}{L_2} \\
\frac{dv_o}{dt} = \frac{i_{L2}}{C_2} - \frac{v_o}{r_L C_2}
\end{cases}
\tag{10.13}
$$

which describes the ideal sliding dynamics of the SM current controlled Ćuk converter.

10.3.5.2 Equilibrium Point

Assume that there exists a stable equilibrium point on the sliding surface on which the ideal sliding dynamics eventually settles. At this point of equilibrium (steady state), there will not be any change in the system's dynamics if there is no input or loading disturbance, i.e., $\frac{di_{L1}}{dt} = \frac{di_{L2}}{dt} = \frac{dv_{C1}}{dt} = \frac{dv_o}{dt} = 0$. Then, the state equations in (10.13) can be equated to give

$$
\begin{cases}
V_{C1} = V_i + V_o \\
I_{L2} = \frac{V_o}{R_L} \\
I_{L1} = I_{L2}\frac{V_o}{V_i} = \frac{V_o^2}{V_i R_L}
\end{cases}
\tag{10.14}
$$

where I_{L1}, I_{L2}, V_{C1}, V_o, V_i, and R_L represents the inductor currents, capacitor and output voltages, input voltage, and load resistance at steady-state equilibrium, respectively.

10.3.5.3 Linearization of Ideal Sliding Dynamics

Next, by separating the ideal sliding dynamic's equation (10.13) into its DC and ac terms, we have

$$
\begin{cases}
\frac{d[I_{L1}+\tilde{i}_{L1}]}{dt} = -\frac{K_3 C_2\frac{d[V_o+\tilde{v}_o]}{dt}+G_s V_i+K_1 K\beta\tilde{v}_o+K_1\tilde{i}_{L1}+K_2 K\beta\int\tilde{v}_o dt+K_2\int\tilde{i}_{L1}dt}{G_s L_1} \\
\qquad\qquad\qquad\qquad\qquad\qquad\qquad\qquad\qquad\qquad\qquad + \frac{V_i}{L_1} \\
\frac{d[V_{C1}+\tilde{v}_{C1}]}{dt} = \frac{K_3 C_2\frac{d[V_o+\tilde{v}_o]}{dt}+G_s V_i+K_1 K\beta\tilde{v}_o+K_1\tilde{i}_{L1}+K_2 K\beta\int\tilde{v}_o dt+K_2\int\tilde{i}_{L1}dt}{G_s V_{C1}} \\
\qquad\qquad \times \frac{I_{L1}+\tilde{i}_{L1}}{C_1} - \frac{I_{L2}+\tilde{i}_{L2}}{C_1} + \frac{I_{L2}+\tilde{i}_{L2}}{C_1}\times \\
\qquad\qquad \frac{K_3 C_2\frac{d[V_o+\tilde{v}_o]}{dt}+G_s V_i+K_1 K\beta\tilde{v}_o+K_1\tilde{i}_{L1}+K_2 K\beta\int\tilde{v}_o dt+K_2\int\tilde{i}_{L1}dt}{G_s V_{C1}} \\
\frac{d[I_{L2}+\tilde{i}_{L2}]}{dt} = \frac{V_{C1}+\tilde{v}_{C1}}{L_2} - \frac{V_o+\tilde{v}_o}{L_2} \\
\qquad\qquad - \frac{K_3 C_2\frac{d[V_o+\tilde{v}_o]}{dt}+G_s V_i+K_1 K\beta\tilde{v}_o+K_1\tilde{i}_{L1}+K_2 K\beta\int\tilde{v}_o dt+K_2\int\tilde{i}_{L1}dt}{G_s L_2} \\
\frac{d[V_o+\tilde{v}_o]}{dt} = \frac{I_{L2}+\tilde{i}_{L2}}{C_2} - \frac{V_o+\tilde{v}_o}{R_L C_2}
\end{cases}
\tag{10.15}
$$

The derivation above is performed assuming the validity of the following steady-state equilibrium conditions, $v_i = V_i$, $r_L = R_L$, $V_{ref} - \beta V_o = 0$, and

$I_{\text{ref}} = I_L = K(V_{\text{ref}} - \beta V_o)$, and the assumptions $K \gg 1$, $V_{C1} \gg \tilde{v}_{C1}$ and $V_o \gg \tilde{v}_o$. Then, by considering only the ac terms, the linearization of the ideal sliding dynamics around the equilibrium point given in (10.14) transforms (10.15) into

$$
\begin{cases}
\dfrac{d\tilde{i}_{L1}}{dt} = -\dfrac{1}{G_s L_1}\left[K_3 \tilde{i}_{L2} + (K_1 K\beta - \dfrac{K_3}{R_L})\tilde{v}_o + K_1 \tilde{i}_{L1} \right. \\
\qquad\qquad \left. + K_2 K\beta \displaystyle\int \tilde{v}_o dt + K_2 \displaystyle\int \tilde{i}_{L1} dt \right] \\[2mm]
\dfrac{d\tilde{v}_{C1}}{dt} = \dfrac{V_o}{V_i R_L C_1 G_s}\left[K_3 \tilde{i}_{L2} + (K_1 K\beta - \dfrac{K_3}{R_L})\tilde{v}_o + K_1 \tilde{i}_{L1} \right. \\
\qquad\qquad \left. + K_2 K\beta \displaystyle\int \tilde{v}_o dt + K_2 \displaystyle\int \tilde{i}_{L1} dt \right] + \dfrac{V_i}{C_1 V_{C1}}\tilde{i}_{L1} - \dfrac{V_o}{C_1[V_i + V_o]}\tilde{i}_{L2} \quad .(10.16) \\[2mm]
\dfrac{d\tilde{i}_{L2}}{dt} = -\dfrac{1}{G_s L_2}\left[K_3 \tilde{i}_{L2} + (K_1 K\beta - \dfrac{K_3}{R_L})\tilde{v}_o + K_1 \tilde{i}_{L1} \right. \\
\qquad\qquad \left. + K_2 K\beta \displaystyle\int \tilde{v}_o dt + K_2 \displaystyle\int \tilde{i}_{L1} dt \right] + \dfrac{1}{L_2}\tilde{v}_{C1} - \dfrac{1}{L_2}\tilde{v}_o \\[2mm]
\dfrac{d\tilde{v}_o}{dt} = \dfrac{1}{C_2}\tilde{i}_{L2} - \dfrac{1}{R_L C_2}\tilde{v}_o
\end{cases}
$$

Then, with equation (10.16) rearranged in the standard form

$$
\begin{cases}
\dfrac{d\tilde{i}_{L1}}{dt} = a_{11}\tilde{i}_{L1} + a_{12}\tilde{v}_{C1} + a_{13}\tilde{i}_{L2} + a_{14}\tilde{v}_o + a_{15}\displaystyle\int \tilde{i}_{L1} dt + a_{16}\displaystyle\int \tilde{v}_o dt \\[2mm]
\dfrac{d\tilde{v}_{C1}}{dt} = a_{21}\tilde{i}_{L1} + a_{22}\tilde{v}_{C1} + a_{23}\tilde{i}_{L2} + a_{24}\tilde{v}_o + a_{25}\displaystyle\int \tilde{i}_{L1} dt + a_{26}\displaystyle\int \tilde{v}_o dt \\[2mm]
\dfrac{d\tilde{i}_{L2}}{dt} = a_{31}\tilde{i}_{L1} + a_{32}\tilde{v}_{C1} + a_{33}\tilde{i}_{L2} + a_{34}\tilde{v}_o + a_{35}\displaystyle\int \tilde{i}_{L1} dt + a_{36}\displaystyle\int \tilde{v}_o dt \\[2mm]
\dfrac{d\tilde{v}_o}{dt} = a_{41}\tilde{i}_{L1} + a_{42}\tilde{v}_{C1} + a_{43}\tilde{i}_{L2} + a_{44}\tilde{v}_o + a_{45}\displaystyle\int \tilde{i}_{L1} dt + a_{46}\displaystyle\int \tilde{v}_o dt \\[2mm]
\dfrac{d[\int \tilde{i}_{L1} dt]}{dt} = a_{51}\tilde{i}_{L1} + a_{52}\tilde{v}_{C1} + a_{53}\tilde{i}_{L2} \\
\qquad\qquad + a_{54}\tilde{v}_o + a_{55}\displaystyle\int \tilde{i}_{L1} dt + a_{56}\displaystyle\int \tilde{v}_o dt \\[2mm]
\dfrac{d[\int \tilde{v}_o dt]}{dt} = a_{61}\tilde{i}_{L1} + a_{62}\tilde{v}_{C1} \\
\qquad\qquad + a_{63}\tilde{i}_{L2} + a_{64}\tilde{v}_o + a_{65}\displaystyle\int \tilde{i}_{L1} dt + a_{66}\displaystyle\int \tilde{v}_o dt
\end{cases}
\tag{10.17}
$$

where

$$
\begin{cases}
a_{11} = -\dfrac{K_1}{G_s L_1}; \quad a_{12} = 0; \quad a_{13} = -\dfrac{K_3}{G_s L_1}; \\[2ex]
a_{14} = \dfrac{K_3}{R_L G_s L_1} - \dfrac{K_1 K}{L_1}; \quad a_{15} = -\dfrac{K_2}{G_s L_1}; \quad a_{16} = -\dfrac{K_2 K}{L_1}; \\[2ex]
a_{21} = \dfrac{K_1 V_o}{V_i R_L C_1 G_s} + \dfrac{V_i}{C_1 V_{C1}}; a_{22} = 0; a_{23} = \dfrac{K_3 V_o}{V_i R_L C_1 G_s} - \dfrac{V_o}{C_1 [V_i + V_o]}; \\[2ex]
a_{24} = \dfrac{(K_1 K R_L G_s - K_3) V_o}{V_i R_L{}^2 C_1 G_s}; a_{25} = \dfrac{K_2 V_o}{V_i R_L C_1 G_s}; a_{26} = \dfrac{K_2 K V_o}{V_i R_L C_1}; \\[2ex]
a_{31} = -\dfrac{K_1}{G_s L_2}; \quad a_{32} = \dfrac{1}{L_2}; \quad a_{33} = -\dfrac{K_3}{G_s L_2}; \\[2ex]
a_{34} = \dfrac{K_3}{R_L G_s L_2} - \dfrac{K_1 K}{L_2}; \quad a_{35} = -\dfrac{K_2}{G_s L_2}; \quad a_{36} = -\dfrac{K_2 K}{L_2}; \\[2ex]
a_{41} = 0; \quad a_{42} = 0; \quad a_{43} = \dfrac{1}{C_2}; \\[2ex]
a_{44} = -\dfrac{1}{R_L C_2}; \quad a_{45} = 0; \quad a_{46} = 0; \\[2ex]
a_{51} = 1; \quad a_{52} = 0; \quad a_{53} = 0; \\[1ex]
a_{54} = 0; \quad a_{55} = 0; \quad a_{56} = 0; \\[1ex]
a_{61} = 0; \quad a_{62} = 0; \quad a_{63} = 1; \\[1ex]
a_{64} = 0; \quad a_{65} = 0; \quad a_{66} = 0,
\end{cases} \tag{10.18}
$$

the characteristic equation of the linearized system will be given by

$$
\begin{vmatrix}
s - a_{11} & 0 & -a_{13} & -a_{14} & -a_{15} & -a_{16} \\
-a_{21} & s & -a_{23} & -a_{24} & -a_{25} & -a_{26} \\
-a_{31} & -a_{32} & s - a_{33} & -a_{34} & -a_{35} & -a_{36} \\
0 & 0 & -a_{43} & s - a_{44} & 0 & 0 \\
-1 & 0 & 0 & 0 & s & 0 \\
0 & 0 & -1 & 0 & 0 & s
\end{vmatrix}
$$
$$
= s^6 + p_1 s^5 + p_2 s^4 + p_3 s^3 + p_4 s^2 + p_5 s + p_6 = 0 \tag{10.19}
$$

where

$$
\begin{cases}
p_1 = -a_{11} - a_{33} - a_{44} \\
p_2 = a_{11}a_{33} + a_{11}a_{44} + a_{33}a_{44} - a_{13}a_{31} \\
\quad\quad - a_{15} - a_{23}a_{32} - a_{34}a_{43} - a_{36} \\
p_3 = a_{11}a_{23}a_{32} + a_{11}a_{34}a_{43} + a_{11}a_{36} + a_{13}a_{31}a_{44} + a_{15}a_{33} \\
\quad\quad + a_{15}a_{44} + a_{23}a_{32}a_{44} + a_{36}a_{44} - a_{11}a_{33}a_{44} - a_{13}a_{21}a_{32} \\
\quad\quad - a_{13}a_{35} - a_{14}a_{31}a_{43} - a_{16}a_{31} - a_{24}a_{32}a_{43} - a_{26}a_{32} \\
p_4 = a_{11}a_{24}a_{32}a_{43} + a_{11}a_{26}a_{32} + a_{13}a_{21}a_{32}a_{44} + a_{13}a_{35}a_{44} \\
\quad\quad + a_{15}a_{23}a_{32} + a_{15}a_{34}a_{43} + a_{15}a_{36} + a_{16}a_{31}a_{44}a_{26}a_{32}a_{44} \\
\quad\quad + a_{35}a_{43} - a_{11}a_{23}a_{32}a_{44} - a_{11}a_{36}a_{44} - a_{13}a_{25}a_{32} \\
\quad\quad - a_{14}a_{21}a_{32}a_{43} - a_{15}a_{33}a_{44} - a_{16}a_{21}a_{32} - a_{16}a_{35} \\
p_5 = a_{13}a_{25}a_{32}a_{44} + a_{15}a_{24}a_{32}a_{43} + a_{15}a_{26}a_{32} + a_{16}a_{21}a_{32}a_{44} \\
\quad\quad + a_{16}a_{35}a_{44} - a_{11}a_{26}a_{32}a_{44} - a_{14}a_{25}a_{32}a_{43} - a_{14}a_{43} \\
\quad\quad - a_{15}a_{23}a_{32}a_{44} - a_{15}a_{36}a_{44} - a_{16}a_{25}a_{32} \\
p_6 = a_{16}a_{25}a_{32}a_{44} - a_{15}a_{26}a_{32}a_{44}
\end{cases}
\tag{10.20}
$$

Application of the Routh-Hurwitz criterion to the characteristic equation in (10.19) concludes that the system will be stable if the following conditions are satisfied:

$$
\begin{cases}
p_1 > 0 \\[4pt]
p_2 > \dfrac{p_3}{p_1} \\[10pt]
p_3 > \dfrac{p_1 p_4 - p_5}{p_2 - \frac{p_3}{p_1}} \\[12pt]
p_4 > \dfrac{p_5}{p_1} + \dfrac{p_5\left(p_2 - \frac{p_3}{p_1}\right) - p_1 p_6}{p_3 - \frac{p_1 p_4 + p_5}{p_2 - \frac{p_3}{p_1}}} \\[16pt]
p_5 > \dfrac{p_1 p_6}{p_2 - \frac{p_3}{p_1}} + \dfrac{p_6\left(p_3 - \frac{p_1 p_4 - p_5}{p_2 - \frac{p_3}{p_1}}\right)}{\dfrac{p_5}{p_1} + \dfrac{p_5\left(p_2 - \frac{p_3}{p_1}\right) - p_1 p_6}{p_3 - \frac{p_1 p_4 + p_5}{p_2 - \frac{p_3}{p_1}}}} \\[20pt]
p_6 > 0
\end{cases}
\tag{10.21}
$$

Hence, by numerically solving equations (10.18) and (10.20), and substituting its solutions in (10.21), the stability of the system can be determined. This, along with the existence condition (10.11), form the basis for the selection and design of the control gains of the controller in conjunction with the specified requirements of the converter.

TABLE 10.1

Specifications of the Ćuk converter.

Description	Parameter	Nominal Value
Input voltage	v_i	24 V
Energy transfer capacitance	C_1	2200 μF
ESR of C_1	r_{C1}	25 mΩ
Output filter capacitance	C_2	230 μF
ESR of C_2	r_{C2}	25 mΩ
Input inductance	L_1	400 μH
Resistance of L_1	r_{l1}	0.12 Ω
Output inductance	L_2	200 μH
Resistance of L_2	r_{l2}	0.12 Ω
Switching frequency	f_S	200 kHz
Minimum load resistance	$r_{L(min)}$	12 Ω
Maximum load resistance (25% of full load)	$r_{L(max)}$	48 Ω
Desired output voltage	V_{od}	36 V

10.3.6 Selection of Sliding Coefficients

The selection of the sliding coefficients follows the approach given in the previous chapter.

10.3.7 Further Comments

Similar constant-frequency reduced-state SM current controllers can be developed for other high-order converters (e.g., Sepic and Zeta converters) using the aforementioned approach. The main point to note when developing these controllers is to choose only the output voltage error and the inductor current error (difference between the current reference profile and the input inductor current) as the controlled state variables to achieve reduced-state control. With that, the same sliding function as in equations (10.3) and (10.4) can be adopted.

10.4 Results and Discussion

Computer simulation[1] and experimental work have been performed on the derived reduced-state PWM-based SM current controller.

Table 10.1 shows the specifications of the Ćuk converter used in the simulation and the experimental prototype. The 24 V input, 100 W converter,

---

[1]The simulation is performed using Matlab/Simulink. The step size taken for all simulations is 10 ns.

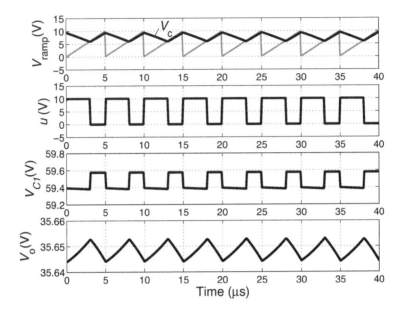

FIGURE 10.3
Waveforms of the control signal v_c, input ramp v_{ramp}, generated gate pulse u, energy transfer capacitor voltage v_{C1}, and output voltage v_o of the PWM-based SM current controlled Ćuk converter operating at $v_i = 24$ V, $v_o = 36$ V, and full-load, i.e., $r_L = 12 \; \Omega$.

is to operate for step-up conversion from an input voltage ranging between 20 V $\leq v_i \leq 28$ V to an output voltage of 36 V. Choosing $V_{ref} = 6$ V, the voltage divider network is set as $\beta = \frac{1}{6}$ V/V. Using the approach mentioned in the previous section, the control parameters of the PWM-based SM current controller are determined as $K = 80$ A/V, $K_1 = 10$ V/A, $K_2 = 5$ V/A, and $K_3 = 1.5$ /A. Thus, the control equation implemented in the simulation is

$$
\begin{cases}
v_c = 10 \left(V_{ref} - \frac{1}{6}v_o \right) + 5 \int \left(V_{ref} - \frac{1}{6}v_o \right) dt + 10 \left(80 \left[V_{ref} - \frac{1}{6}v_o \right] - i_{L1} \right) \\
\quad + 5 \int \left(80 \left[V_{ref} - \frac{1}{6}v_o \right] - i_{L1} \right) dt - 1.5 i_{C2} + \frac{1}{6} \left(v_{C1} - v_i \right) \\
\hat{v}_{ramp} = \frac{1}{6} v_{C1}
\end{cases}
\tag{10.22}
$$

10.4.1 Steady-State Performance

Figure 10.3 shows the steady-state waveforms of the SM current controlled Ćuk converter operating at nominal input voltage and full-load condition. The control signal v_c is generated from analog computation of the instantaneous feedback signals v_o, v_i, v_{C1}, i_{L1}, and i_{C2} using the expression described in (10.22). Also, the peak amplitude of the ramp signal v_{ramp} is generated to

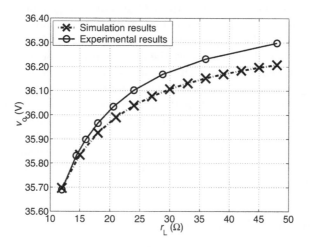

(a) Load regulation property

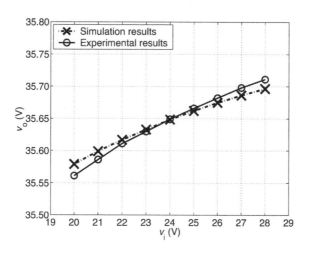

(b) Line regulation property

FIGURE 10.4
Plot of DC output voltage v_o against (a) load resistance r_L at nominal input voltage $v_i = 24$ V; and (b) input voltage v_i at full-load condition $r_L = 12\ \Omega$.

follow $\frac{1}{6} v_{C1}$ as in (10.22). Both v_c and v_{ramp} are fed into the pulse-width modulator to generate the gate pulse u for the switching of the Ćuk converter. The corresponding voltage waveforms of the energy transfer and output capacitors are also illustrated in the figure.

Figure 10.4(a) shows a plot of the DC output voltage against the different operating load resistances at nominal input voltage condition. The result shows good load regulation property for the load range of $12 \ \Omega \leq r_{\mathrm{L}} \leq 48 \ \Omega$ with only a 0.26 V deviation in v_{o} (i.e., about 0.72 % of $v_{\mathrm{o}(24 \ \mathrm{V};12 \ \Omega)}$) for the computer simulation, and with only a 0.30 V deviation in v_{o} (i.e., about 0.84 % of $v_{\mathrm{o}(24 \ \mathrm{V};12 \ \Omega)}$) for the experimental measurement.

Figure 10.4(b) shows a plot of the DC output voltage against the input voltage at full-load condition. The result shows very good line regulation property for the input range of $20 \ \mathrm{V} \leq v_{\mathrm{i}} \leq 28 \ \mathrm{V}$ with only a 0.12 V deviation in v_{o} (i.e., about 0.33 % of $v_{\mathrm{o}(24 \ \mathrm{V};12 \ \Omega)}$) for the computer simulation, and with only a 0.15 V deviation in v_{o} (i.e., about 0.42 % of $v_{\mathrm{o}(24 \ \mathrm{V};12 \ \Omega)}$) for the experimental measurement.

10.4.2   Dynamic Performance

Figures 10.5 and 10.6 show the input/output current and voltage waveforms of the converter at dynamic load condition that are obtained respectively from simulation and experimental measurement. It is shown that the reduced-state PWM-based SM current controller is capable of achieving good dynamical response to the Ćuk converter, with a settling time of around 15 ms with small overshoots, when the output power is stepped between 27 W (quarter load) and 108 W (full load). Moreover, the perfect consistency in the dynamical behavior of the output voltage response when the load resistance steps up and down clearly demonstrates the superiority of the controller for large-signal operation.

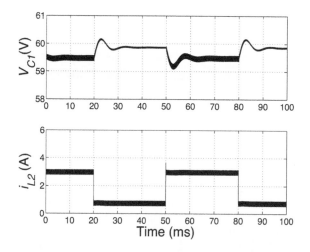

(a) v_{C1} (1 V/DIV); i_{L2} (1 A/DIV)

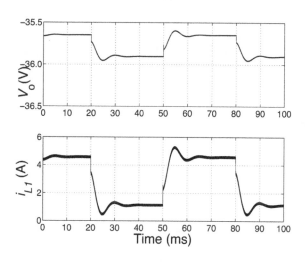

(b) v_o (0.5 V/DIV); i_{L1} (2 A/DIV)

FIGURE 10.5
Simulation waveforms of input filter current i_{L1}, output filter current i_{L2}, energy transfer capacitor voltage v_{C1}, and output voltage v_o of the PWM-based SM current controlled Ćuk converter operating at $v_i = 24$ V and load resistance alternating between full load and quarter load, i.e., $r_L = 12\ \Omega$ and $48\ \Omega$.

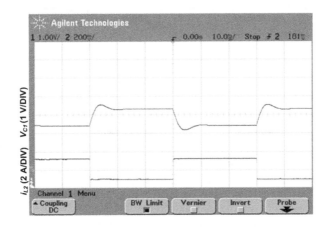

(a) v_{C1} (1 V/DIV); i_{L2} (1 A/DIV)

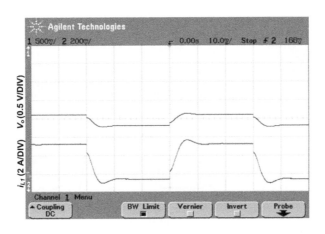

(b) v_o (0.5 V/DIV); i_{L1} (2 A/DIV)

FIGURE 10.6
Experimental waveforms of input filter current i_{L1}, output filter current i_{L2}, energy transfer capacitor voltage v_{C1}, and output voltage v_o of the PWM-based SM current controlled Ćuk converter operating at $v_i = 24$ V and load resistance alternating between full load and quarter load, i.e., $r_L = 12\ \Omega$ and $48\ \Omega$. Note that the measurements have been carried out with high-frequency filters that remove the switching ripples.

11

Indirect Sliding Mode Control with Double Integral Sliding Surface

CONTENTS

11.1   Introduction

It should be quite clear by now that the actual application of SM controllers in power converters have been hindered by two major challenges: the varying operating frequency of the SM controller, and the presence of non-negligible steady-state error in the regulation due to practical non-idealities. Regarding the first issue, several methods of fixing the switching frequency, which include the incorporation of constant timing functions or circuitries into the

SM controllers [9, 35, 52], the use of adaptive strategies as discussed in Chapter 5, and the indirect implementation of the SM controllers using the PWM approach as extensively elaborated in Chapters 6 to 10, have been proposed. As for the second issue, it has been widely known that the steady-state error of an SM-controlled system can be effectively suppressed through the use of an additional integral term of the state variables in the SM controller [11, 23, 52, 55, 63, 84]. This method is known as *integral sliding mode control.* When incorporated, the consequence is an SM-controlled system 1) with motion equation of the same order as the original system; and 2) with improved robustness and regulation property than the traditional SM-controlled system [102].

However, the adoption of the integral sliding mode (ISM) control scheme in these controllers *cannot effectively eliminate* the steady-state error of the converters. Our investigation shows that this is due to the imperfect steady-state error correction method of the pulse-width modulation (PWM)-based ISM controllers. The problem is common to all types of indirect ISM controllers derived from the equivalent control method [44, 45, 63, 90, 93, 96]. It is also found that the steady-state error increases as the converter's switching frequency decreases and that the error can be quite large at a low switching frequency.

Since increasing the order of the controller of a system generally improves the steady-state accuracy [65], it is possible to eliminate the steady-state regulation error of the indirect ISM-controlled converter by increasing the order of the indirect ISM controller using an additional integral term. This chapter gives an in-depth discussion of the phenomenon and also reports the effectiveness of incorporating double-integral state variables in constructing the sliding manifold of indirect SM controllers for power converters for suppressing the steady-state error. Various aspects of the solution including its design and implementation are discussed in terms of the fixed-frequency PWM-based (indirect) SM controller. While the discussion in this chapter is carried out in the context of power converters, the basic technique is applicable to other types of systems.

11.2 Problem Identification

11.2.1 Review of Hysteresis-Modulation-Based Sliding Mode Controllers

Conventional *direct* SM controllers based on hysteresis-modulation (HM) are implemented through the real-time computation of the state variables to generate a suitable profile of the system trajectory, which moves in the proximity of a desired sliding surface toward the equilibrium state. Since SM control can achieve order reduction, it is typically sufficient to have an SM controller of $(n-1)$th order for achieving stable control of an nth order converter. For instance, the switching function of a common form of the SM controller for an nth order converter is given by

$$u = \begin{cases} u^+ & \text{when } S > \kappa \\ u^- & \text{when } S < -\kappa \end{cases} \tag{11.1}$$

where κ is a parameter controlling the switching frequency of the system, and S is the instantaneous system trajectory of reduced order, which is expressed as

$$S = \sum_{i=1}^{n-1} \alpha_i x_i, \tag{11.2}$$

where α_i for $i = 1$ to $n-1$ denotes the sets of control parameters, i.e., sliding coefficients. Under the configuration $\kappa = 0$, it is typically assumed in SM control that the controller/converter system operates ideally at an infinite switching frequency with no steady-state error.

However, practical non-ideality limits the switching frequency to be a finite quantity which deteriorates the robustness and regulation properties of the converter system controlled by the reduced-order SM controller. Steady-state errors are therefore present. This can be observed from Fig. 11.1, which illustrates the regulation property[1] of a 12 V buck converter using the HM-based SM controller, the HM-based ISM controller, and the PWM-based (indirect) ISM controller, at various switching frequencies. The plot also reflects the kind of behavior expected in a converter with the HM-based SM controller, that is, the steady-state error increases as the switching frequency reduces.

A good method of suppressing these errors is to introduce an additional integral term of the state variables to the SM controller, which transforms it into an ISM controller. Since the ISM controller is of the same order as the converter, it is also known as a *full-order SM controller* [102]. Such an ISM controller can be obtained by modifying (11.2) into

$$S = \sum_{i=1}^{n-1} \alpha_i x_i + \alpha_n \int \sum_{i=1}^{n-1} x_i dt, \tag{11.3}$$

where the additional state variable (as compared to (11.2)) is basically the integral term of all other existing state variables. Notably, it is the component $\int x_i dt$ which *directly* nullifies the steady-state errors of the respective state variables, i.e., x_i. The effect of such property can be seen in Fig. 11.1. It can be seen that with the HM-based ISM controller, the output voltage of the

---

[1]The presented results are obtained from computer simulation to avoid discrepancy arising from the variation of the experimental setups. They basically reflect the same characteristics as the actual experimental data.

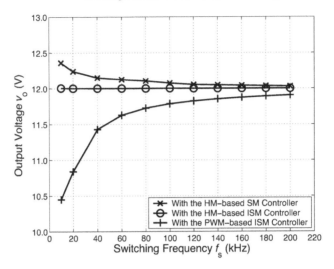

FIGURE 11.1
Plot of steady-state output voltage v_o against switching frequency f_s of a 12 V output buck converter operating with 3 Ω load under HM-based SM controller with and without integral control.

converter is maintained at 12 V with a negligible steady-state error at all switching frequencies.

However, when it comes to the indirect implementation of the ISM controller, the effectiveness of the integral control in eliminating the steady-state error deteriorates. The plot of the PWM-based ISM controller in Fig. 11.1 clearly demonstrates such an outcome. The following section discusses why the ISM controller succeeds in eliminating the steady-state error in the direct (HM) form, but fails in the indirect (PWM) form.

11.2.2   Review of Indirect Sliding Mode Controllers

Firstly, for implementation of any SM controller in the indirect form, the original control law must be expressed in an alternate form. This is based on the *equivalent control* method, which assumes the invariance conditions that during SM operation, $S = 0$ and $\dot{S} = 0$. From such an assumption, an equivalent control signal u_{eq} can be derived in terms of the respective state variables. Hence, the trajectory S is indirectly formulated to track the desired sliding surface through the construction of the control signal u_{eq}. This makes it an *indirect* approach for ensuring SM operation. Note that similar to the direct SM control approach, the indirect approach must include the hitting condition which ensures the state trajectory being driven toward the sliding surface and the existence condition which ensures that the state trajectory is kept within the vicinity of the surface. Only with such constraints will the equivalent control signal derived from the original control law ensures SM operation.

To derive the equivalent control, the time differentiation of (11.3) is first derived, i.e.,

$$\dot{S} = \sum_{i=1}^{n-1} \alpha_i \dot{x}_i + \alpha_n \sum_{i=1}^{n-1} x_i. \tag{11.4}$$

Equating $\dot{S} = 0$ and solving for u_{eq} give the general form

$$u_{eq} = G(\dot{x}_1, \dot{x}_2, ..., \dot{x}_{n-1}, x_1, x_2, ..., x_{n-1}) \tag{11.5}$$

where $0 < u_{eq} < 1$ is a function of the state variables $\dot{x}_i$ and x_i for $i = 1, 2, ..., n - 1$. In practice, in the case of PWM-based (indirect) SM controller implementation, the control signal u_{eq} is constructed through a pulse-width modulator using a constant frequency ramp signal v_{ramp} and a feedback control signal v_c, where $u_{eq} = \frac{v_c}{v_{ramp}}$. Hence, both v_{ramp} and v_c are functions of the state variables $\dot{x}_i$ and x_i. It is important to pinpoint that the indirect construction of S using the indirect approach (such as PWM) uses state variables of one time derivative order lower than the original HM-based ISM controller (see equation (11.3)). This explains why the steady-state error correction succeeds in HM-based ISM controller but fails in the indirect ISM controller, and why the problem is particularly severe when the switching frequency is low.

11.2.3 Analytical Explanation for the Presence of Steady-State Error in Indirect ISM-Controlled Converter

In the case of the direct (HM-based) ISM controller, the sliding surface constructed comprises the integral elements of the steady-state errors, i.e., $\int x_i dt$ for $i = 1, 2, ..., n - 1$. Recall that $\int x_i dt$ is a component that directly accumulates the existing steady-state errors. Hence, when the state trajectory S is directed to move on the sliding surface toward an equilibrium point, the steady-state error is automatically eliminated. With this process of closed-loop steady-state-error-correction feedback, the switching frequency will have little influence on the magnitude of the steady-state error present in the HM-based ISM controlled converter.

However, for the indirect ISM controller, the variables $\int x_i dt$ are not explicitly reflected in the control signal (see equation (11.5)). Instead, these integral functions are embedded in the sliding surface and the required error corrections are indirectly computed using the state variables $\dot{x}_i$ and x_i. Since there is no direct integral signal $\int x_i dt$ that corrects the errors of the state variables, the capability of the correction is then dependent on the accuracy of the *indirect integral computation*. However, such computations are open-loop processes which contain finite steady-state errors that cannot be eliminated. Hence, with steady-state errors present in the computation, steady-state errors will be present in the controlled variables. Naturally, this problem will be further aggravated if the switching frequency reduces. It is, therefore, an inefficient method of steady-state error correction and is an obvious drawback of the indirect SM controller. This explains why the integral control scheme of the indirect ISM controller is ineffective in eliminating the steady-state error[2], especially in the low frequency range.

11.3   A Possible Solution

It is well known that the increased order of the controller improves the steady-state accuracy of the system, but aggravates the stability problem [65]. An additional double-integral term of the state variables, i.e., $\int \left[\int x_i dt \right] dt$ for $i = 1, 2, ..., n-1$, is therefore introduced to *correct the error of the indirect integral computation* in the indirect ISM controller. By using an integral closed-loop to eliminate the steady-state error of the indirect integral computation, the steady-state errors of the controlled state variables can be eliminated. This is the so-called double-integral (indirect) sliding mode (DISM) controller discussed in this chapter. Notably, the solution is simple and straightforward. But its use in practical design is rarely reported and requires investigation and validation.

In its general direct HM form, the DISM controller takes the switching function shown in (11.1), where

$$S = \sum_{i=1}^{n-1} \alpha_i x_i + \alpha_n \int \sum_{i=1}^{n-1} x_i dt + \alpha_{n+1} \int \int \sum_{i=1}^{n-1} x_i dt dt. \qquad (11.6)$$

Clearly, this controller is one order higher than the original converter system.[3] Its time differentiation

$$\dot{S} = \sum_{i=1}^{n-1} \alpha_i \dot{x}_i + \alpha_n \sum_{i=1}^{n-1} x_i + \alpha_{n+1} \int \sum_{i=1}^{n-1} x_i dt \qquad (11.7)$$

is likewise an order higher than the ISM (full-order) controller represented in (11.4). By solving $\dot{S} = 0$, it is not difficult to see that the equivalent control u_{eq}

---

[2]In practice, it is possible to up/down lift the reference setpoint to obtain the desired output. However, it should be noted that such a method of correction *does not* actually eliminate the steady-state errors. For operating conditions deviated from the desired point, the variation of the steady-state errors will deteriorate the regulation of the converter. To achieve a control output that follows closely the reference setpoint for all operating conditions, the feedback loop is still required to suppress the steady-state errors for the entire range of operating conditions.

[3]Note that the order of the controller does not correspond to the number of state variables in the controller. Even though $n-1$ state variables have been added, the actual order of the controller increases by only one.

is a function G of the state variables $\dot{x}_i$, x_i, and $\int x_i dt$. Here, the additional term $\int x_i dt$ (as compared to ISM controller) is resulted from the double-integral term $\int \left[\int x_i dt \right] dt$ introduced by the DISM controller. It is interesting to see that by directly correcting the steady-state error in x_i, the original objective of introducing this component to *correct the error of the indirect integral computation so that the steady-state errors of the controlled variables are eliminated* is inherently met. The DISM configuration easily resolves the problem of steady-state errors in indirect ISM-controlled converters.

11.4 Application of Double-Integral Sliding Surface to PWM-Based Types of Indirect Sliding Mode Controllers

This section discusses the application of the afore-described solution, i.e., DISM configuration, to the PWM-based SM controller for the voltage controlled buck converter discussed in Chapter 8 and the current controlled boost converter discussed in Chapter 9.

11.4.1 Double-Integral Sliding Mode Controllers

In the examples of the DISM controller for buck converters and boost converters, we use the switching function $u = \frac{1}{2}(1 + \mathrm{sign}(S))$ and the sliding surface

$$S = \alpha_1 x_1 + \alpha_2 x_2 + \alpha_3 x_3 + \alpha_4 x_4 \tag{11.8}$$

where u represents the logic state of power switch S_W, and α_1, α_2, α_3, and α_4 represent the desired sliding coefficients. Also, in both examples, C, L, and r_L denote the capacitance, inductance, and instantaneous load resistance respectively; V_ref, v_i, and v_o denote the reference, instantaneous input, and instantaneous output voltages respectively; β denotes the feedback network ratio; i_ref, i_L, i_C, and i_r denote the instantaneous reference, instantaneous inductor, instantaneous capacitor, and instantaneous output currents, respectively; and $\bar{u} = 1 - u$ is the inverse logic of u.

Case 1—Buck Converter

For the DISM voltage controlled buck converter, the controlled state variables are the *voltage error* x_1, the *voltage error dynamics* (or the rate of change of voltage error) x_2, the *integral of voltage error* x_3, and the *double integral of the voltage error* x_4, which are expressed as

$$
\begin{cases}
x_1 = V_{ref} - \beta v_o \\
x_2 = \dot{x}_1 \\
x_3 = \int x_1 dt \\
x_4 = \int \left(\int x_1 dt \right) dt
\end{cases}
\tag{11.9}
$$

Substitution of the buck converter's behavioral model under continuous conduction mode of operation into the time differentiation of (11.9) gives the dynamical model of the system as

$$
\begin{cases}
\dot{x}_1 = \dfrac{d[V_{ref} - \beta v_o]}{dt} = -\dfrac{\beta}{C} i_C \\[2mm]
\dot{x}_2 = \dfrac{\beta}{r_L C^2} i_C - \dfrac{\beta v_i}{LC} u + \dfrac{\beta v_o}{LC} \\[2mm]
\dot{x}_3 = x_1 = V_{ref} - \beta v_o \\[2mm]
\dot{x}_4 = \int x_1 dt = \int (V_{ref} - \beta v_o) dt
\end{cases}
\tag{11.10}
$$

The equivalent control signal of the DISM voltage controller when applied to the buck converter is obtained by solving $\frac{dS}{dt} = 0$, giving

$$
\begin{aligned}
u_{eq} &= \frac{\beta L}{\beta v_i} \left(\frac{1}{r_L C} - \frac{\alpha_1}{\alpha_2} \right) i_C + \frac{\beta v_o}{\beta v_i} \\
&+ \frac{\alpha_3}{\alpha_2} \frac{LC}{\beta v_i} (V_{ref} - \beta v_o) + \frac{\alpha_4}{\alpha_2} \frac{LC}{\beta v_i} \int (V_{ref} - \beta v_o) \, dt
\end{aligned}
\tag{11.11}
$$

where u_{eq} is continuous and bounded by 0 and 1, i.e., $0 < u_{eq} < 1$.

Case 2—Boost Converter

For the DISM current controlled boost converter, the controlled state variables are the *current error* x_1, the *voltage error* x_2, the *integral of the sum of the current and the voltage errors* x_3, and the *double integral of the sum of the current and the voltage errors* x_4, which are expressed as

$$
\begin{cases}
x_1 = i_{ref} - i_L \\
x_2 = V_{ref} - \beta v_o \\
x_3 = \int [x_1 + x_2] \, dt \\
x_4 = \int (\int [x_1 + x_2] \, dt) dt
\end{cases}
\tag{11.12}
$$

where

$$
i_{ref} = K [V_{ref} - \beta v_o]
\tag{11.13}
$$

and K is the amplified gain of the voltage error. Substituting the boost converter's behavioral model under continuous conduction mode into the time differentiation of (11.12) gives the dynamical model of the system as

$$
\begin{cases}
\dot{x}_1 = \dfrac{d[i_{\text{ref}} - i_L]}{dt} = -\dfrac{\beta K}{C} i_C - \dfrac{v_i - \bar{u} v_o}{L} \\[2ex]
\dot{x}_2 = \dfrac{d[V_{\text{ref}} - \beta v_o]}{dt} = -\dfrac{\beta}{C} i_C \\[2ex]
\dot{x}_3 = x_1 + x_2 \\[1ex]
\dot{x}_4 = \int [x_1 + x_2]\, dt
\end{cases}
\tag{11.14}
$$

Using the same approach, the equivalent control signal of the DISM current controller for the boost converter is derived as

$$
u_{\text{eq}} = 1 - \frac{\beta L}{C v_o}\left(K + \frac{\alpha_2}{\alpha_1} \right) i_C - \frac{v_i}{v_o} + \frac{\alpha_3 L}{\alpha_1 v_o}\left(V_{\text{ref}} - \beta v_o \right)
$$

$$
+ \frac{\alpha_3 L}{\alpha_1 v_o}\left[K\left(V_{\text{ref}} - \beta v_o \right) - i_L \right] + \frac{\alpha_4 L}{\alpha_1 v_o} \int \left(V_{\text{ref}} - \beta v_o \right) dt
$$

$$
+ \frac{\alpha_4 L}{\alpha_1 v_o} \int \left[K\left(V_{\text{ref}} - \beta v_o \right) - i_L \right] dt
\tag{11.15}
$$

where u_{eq} is continuous and bounded by 0 and 1.

11.4.2 Architecture of DISM Controllers in PWM Form

Case 1—Buck Converter

In PWM form, the DISM voltage controller for the buck converter inherits the expression

$$
\begin{cases}
v_c = -K_1 i_C + K_2 \left(V_{\text{ref}} - \beta v_o \right) + K_3 \int \left(V_{\text{ref}} - \beta v_o \right) dt + \beta v_o \\[1.5ex]
\hat{v}_{\text{ramp}} = \beta v_i
\end{cases}
\tag{11.16}
$$

where

$$
K_1 = \beta L \left(\frac{\alpha_1}{\alpha_2} - \frac{1}{r_L C} \right); \quad K_2 = \frac{\alpha_3}{\alpha_2} LC; \quad \text{and} \quad K_3 = \frac{\alpha_4}{\alpha_2} LC
\tag{11.17}
$$

are the fixed gain parameters in the controller. The method of deriving the control law (11.16) from the equivalent control expression (11.11) follows the procedure detailed in Chapter 6. Figure 11.2 shows a schematic diagram of the derived PWM-based DISM voltage controller for the buck converter.

Case 2—Boost Converter

In PWM form, the DISM current controller for the boost converter inherits the expression

$$
\begin{cases}
v_c = G_s \left(v_o - v_i \right) + K_1 \left(V_{\text{ref}} - \beta v_o \right) + K_2 \int \left(V_{\text{ref}} - \beta v_o \right) dt - K_3 i_C \\[1.5ex]
\quad + K_1 \left[K\left(V_{\text{ref}} - \beta v_o \right) - i_L \right] + K_2 \int \left[K\left(V_{\text{ref}} - \beta v_o \right) - i_L \right] dt \\[1.5ex]
\hat{v}_{\text{ramp}} = G_s v_o
\end{cases}
\tag{11.18}
$$

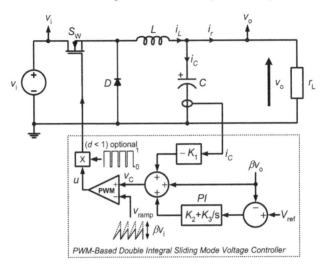

FIGURE 11.2
PWM-based sliding mode voltage controller for buck converters.

where

$$K_1 = G_s \frac{\alpha_3 L}{\alpha_1}; \quad K_2 = G_s \frac{\alpha_4 L}{\alpha_1}; \quad \text{and} \quad K_3 = G_s \frac{\beta L}{C} \left(K + \frac{\alpha_2}{\alpha_1} \right) \quad (11.19)$$

are the fixed gain parameters in the controller. A factor of $0 < G_s < 1$ has been introduced to scale down the equation to conform to the chip level voltage standard. Assuming $\beta = G_s$, the analog implementation of the derived PWM-based DISM current controller for the boost converters is illustrated in Fig. 11.3.

Remarks

The only physical difference between the PWM-based DISM controller and the PWM-based ISM controller is the additional integral terms, i.e., $K_3 \int (V_{\text{ref}} - \beta v_o) \, dt$ and $K_2 \int [K (V_{\text{ref}} - \beta v_o) - i_L] \, dt$. Note that in analog implementation, the proportional (P) and integral (I) components can be easily grouped into a single proportional-integral (PI) type computation. Hence, the complexity of the implementation of the PWM-based ISM controller is similar to that of the PWM-based DISM controller.

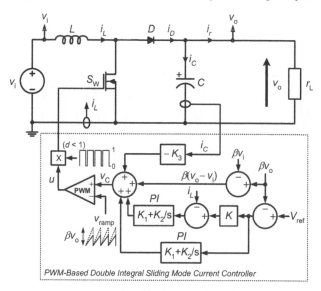

FIGURE 11.3
PWM-based sliding mode current controller for boost converters.

11.4.3 Existence Condition

To ensure the existence of SM operation, the local reachability condition $\lim_{S \to 0} S \cdot \dot{S} < 0$ must be satisfied. This can be expressed as

$$
\begin{cases}
\dot{S}_{S \to 0^+} < 0 \\
\dot{S}_{S \to 0^-} > 0
\end{cases}
. \tag{11.20}
$$

Case 1—Buck Converter

For the DISM voltage controlled buck converter, the existence condition for steady-state operation (equilibrium point) can be derived by substituting (11.8) and its time derivative into (11.20), with consideration of the system's dynamics (11.10), as

$$
\begin{cases}
-K_1 i_{C(\min)} + K_2 x_{1(\max)} + K_3 x_{3(\max)} < \beta \left(v_{i(\min)} - v_{o(ss)} \right) \\
K_1 i_{C(\max)} - K_2 x_{1(\min)} - K_3 x_{3(\min)} < \beta v_{o(ss)}
\end{cases}
, \tag{11.21}
$$

where $v_{i(\min)}$ denotes the minimum input voltage; $v_{o(ss)}$ denotes the expected steady-state output, i.e., approximately the desired reference voltage V_{ref}; $i_{C(\max)}$ and $i_{C(\min)}$ are respectively the maximum and minimum capacitor currents at full-load condition; $x_{1(\max)}$ and $x_{1(\min)}$ are respectively the maximum and minimum steady-state voltage errors, which in this case are basically the inverse functions of the output voltage ripples; and $x_{3(\max)}$ and $x_{3(\min)}$ are respectively the maximum and minimum integrals of the steady-state voltage

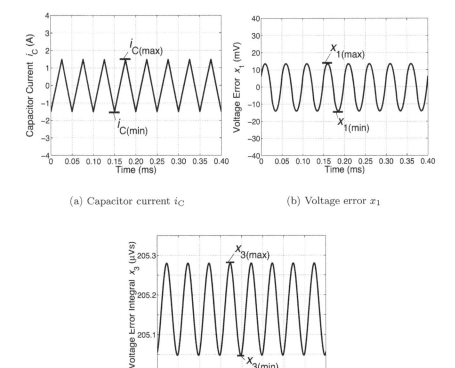

(a) Capacitor current i_C

(b) Voltage error x_1

(c) Integral of voltage error x_4

FIGURE 11.4
Steady-state waveforms of the various state variables of the buck converter at minimum input voltage and full-load condition under ideal open-loop control with negligible output voltage steady-state error.

error, which are the time integrals of the inverse functions of the output voltage ripples with a negligible DC shift. All these parameters can be calculated from the design specifications of the converter. Alternatively, computer simulation of the converter under an ideal open-loop control which gives negligible output voltage steady-state error can be performed to obtain this value. An example of this is given in Fig. 11.4, which shows the waveforms and magnitudes of the various state variables under minimum input voltage and full-load operating condition. Thus, the compliance of the inequalities in (11.21) through the substitution of the state variables' parameters assures the existence of the SM operation to occur at least in the small region of the origin for the designated input and output operating conditions.

Case 2—Boost Converter

For the DISM current controlled boost converter, the existence condition for steady-state operations can be derived by substituting (11.8) and its time derivative into (11.20), with consideration of the system's dynamics (11.14), as

$$\begin{cases} -K_3 i_{C(\min)} + K_1 \left(x_{1(\max)} + x_{2(\max)} \right) + K_2 x_{3(\max)} < \beta v_{i(\min)} \\ K_3 i_{C(\max)} - K_1 \left(x_{1(\min)} + x_{2(\min)} \right) - K_2 x_{3(\min)} < \beta \left(v_{o(ss)} - v_{i(\max)} \right) \end{cases} \tag{11.22}$$

where $v_{i(\max)}$ and $v_{i(\min)}$ denote the maximum and minimum input voltages respectively; $v_{o(SS)}$ denotes the expected steady-state output, i.e., approximately the desired reference voltage V_{ref}; and $i_{L(\max)}$, $i_{L(\min)}$, $i_{C(\max)}$, and $i_{C(\min)}$ are respectively the maximum and minimum inductor and capacitor currents at full-load condition; $x_{1(\max)}$ and $x_{1(\min)}$ are respectively the maximum and minimum steady-state current errors; $x_{2(\max)}$ and $x_{2(\min)}$ are respectively the maximum and minimum steady-state voltage errors, which in this case are basically the inverse functions of the output voltage ripples; and $x_{3(\max)}$ and $x_{3(\min)}$ are respectively the maximum and minimum integrals of the combination of steady-state voltage and current errors. Figure 11.5 illustrates the physical representations of these parameters. Likewise, the design of control parameters must satisfy the inequalities shown in (11.22).

Remarks

It is worth mentioning that compliance with the existence condition, which takes into consideration the amount of ac perturbation of the state variables, indirectly counterchecks the minimum allowable switching frequency for a particular control parameter design, i.e., controller bandwidth. This is similar to the conventional controller design approach in which, as a rule of thumb, the crossover frequency of the controller must be kept at least below one quarter of the switching frequency in the case of a buck converter.

11.4.4 Stability Condition

As in any SM controlled system, the converter under the DISM controller is stable if the real parts of all the *eigenvalues* of the *Jacobian* of the system are negative.

Case 1—Buck Converter

In the DISM voltage controlled buck converter, where the SM equation is the linear motion equation of the converter in SM operation, the stability condition can easily be obtained analytically. This is possible by substituting $S = 0$ into (11.8), i.e.,

$$\alpha_1 x_1 + \alpha_2 x_2 + \alpha_3 x_3 + \alpha_4 x_4 = 0 \tag{11.23}$$

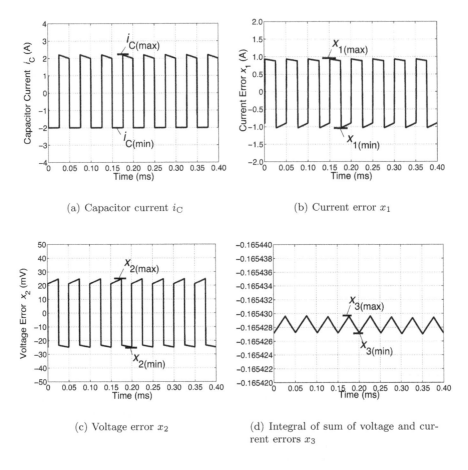

(a) Capacitor current i_C

(b) Current error x_1

(c) Voltage error x_2

(d) Integral of sum of voltage and current errors x_3

FIGURE 11.5
Steady-state waveforms of the various state variables of the boost converter at minimum input voltage and full-load condition under ideal open-loop control with negligible output voltage steady-state error.

to first obtain the motion equation. Since the state variables x_1, x_2, x_3, and x_4 are in phase canonical form, (11.23) can be rewritten in Laplace form as

$$\alpha_1 X(s) + \alpha_2 s X(s) + \alpha_3 \frac{X(s)}{s} + \alpha_4 \frac{X(s)}{s^2} = 0$$

$$\Rightarrow \quad s^3 + \frac{\alpha_1}{\alpha_2} s^2 + \frac{\alpha_3}{\alpha_2} s + \frac{\alpha_4}{\alpha_2} = 0. \qquad (11.24)$$

Finally, applying the Routh-Hurwitz stability criterion to this third-order linear polynomial, we conclude that all coefficients must be positive, i.e., $0 < \alpha_{n=1,2,3,4}$, and that $\alpha_1\alpha_3 > \alpha_2\alpha_4$ to ensure that all roots have negative real parts. On the other hand, the stability condition can be automatically satisfied by designing the sliding coefficients for the desired dynamic response.

Case 2—Boost Converter

For the DISM current controlled boost converter, the motion equation is non-linear. As discussed in Chapters 9 and 10, a different approach based on the *equivalent control method* is required for finding the stability condition. This involves first obtaining the *ideal sliding dynamics* of the system, and then doing a stability analysis on its *equilibrium point*.

11.4.4.1 Ideal Sliding Dynamics

The replacement of $\bar{u}$ by $\bar{u}_{\text{eq}}$ in the original boost converter's description under continuous conduction mode operation converts the discontinuous system into an ideal SM continuous system:

$$\begin{cases} \dfrac{di_L}{dt} = \dfrac{v_{\text{i}}}{L} - \dfrac{v_{\text{o}}}{L}\bar{u}_{\text{eq}} \\ \dfrac{dv_{\text{o}}}{dt} = \dfrac{i_L}{C}\bar{u}_{\text{eq}} - \dfrac{v_{\text{o}}}{r_{\text{L}}C} \end{cases}. \tag{11.25}$$

Then, the substitution of the equivalent control signal into (11.25) gives (11.26),

$$\begin{cases} \frac{di_L}{dt} = \frac{v_{\text{i}}}{L} - \frac{v_{\text{o}}}{L} \times \\ \qquad \frac{K_1\frac{v_{\text{o}}}{r_{\text{L}}}-v_{\text{i}}+K_2[V_{\text{ref}}-\beta v_{\text{o}}]+K_2[i_{\text{ref}}-i_L]+K_3\int[V_{\text{ref}}-\beta v_{\text{o}}]dt+K_3\int[i_{\text{ref}}-i_L]dt}{K_1 i_L - v_{\text{o}}} \\ \frac{dv_{\text{o}}}{dt} = -\frac{v_{\text{o}}}{r_{\text{L}}C} + \frac{i_L}{C} \times \\ \qquad \frac{K_1\frac{v_{\text{o}}}{r_{\text{L}}}-v_{\text{i}}+K_2[V_{\text{ref}}-\beta v_{\text{o}}]+K_2(i_{\text{ref}}-i_L)+K_3\int[V_{\text{ref}}-\beta v_{\text{o}}]dt+K_3\int[i_{\text{ref}}-i_L]dt}{K_1 i_L - v_{\text{o}}} \end{cases} \tag{11.26}$$

which represents the ideal sliding dynamics of the SM current controlled boost converter.

11.4.4.2 Equilibrium Point

Assume that there exists a stable equilibrium point on the sliding surface on which the ideal sliding dynamics eventually settled. Setting all the time derivatives to 0, we get the steady-state equation as

$$I_L = \frac{V_{\text{o}}^2}{V_{\text{i}}R_{\text{L}}} \tag{11.27}$$

where I_L, V_{o} and V_{i} represents the steady-state inductor current, output voltage, and input voltage, respectively.

11.4.4.3 Linearization of Ideal Sliding Dynamics

Next, the linearization of the ideal sliding dynamics around the equilibrium point gives

$$\begin{cases} \dfrac{d\tilde{i}_L}{dt} = a_{11}\tilde{i}_L + a_{12}\tilde{v}_o + a_{13}\displaystyle\int \tilde{i}_L dt + a_{14}\displaystyle\int \tilde{v}_o dt \\[2mm] \dfrac{d\tilde{v}_o}{dt} = a_{21}\tilde{i}_L + a_{22}\tilde{v}_o + a_{23}\displaystyle\int \tilde{i}_L dt + a_{24}\displaystyle\int \tilde{v}_o dt \\[2mm] \dfrac{d[\int \tilde{i}_L dt]}{dt} = a_{31}\tilde{i}_L + a_{32}\tilde{v}_o + a_{33}\displaystyle\int \tilde{i}_L dt + a_{34}\displaystyle\int \tilde{v}_o dt \\[2mm] \dfrac{d[\int \tilde{v}_o dt]}{dt} = a_{41}\tilde{i}_L + a_{42}\tilde{v}_o + a_{43}\displaystyle\int \tilde{i}_L dt + a_{44}\displaystyle\int \tilde{v}_o dt \end{cases} \tag{11.28}$$

where

$$\begin{cases} a_{11} = \dfrac{K_2 V_i R_L}{K_1 L V_o - L V_i R_L}; & a_{12} = -\dfrac{K_1 2V_i - \frac{V_i^2 R_L}{V_o} - K_2 K \beta \frac{V_i}{R_L}}{K_1 L V_o - L V_i R_L}; \\[3mm] a_{13} = \dfrac{K_3 V_i R_L}{K_1 L V_o - L V_i R_L}; & a_{14} = \dfrac{K_3 K \beta \frac{V_i}{R_L}}{K_1 L V_o - L V_i R_L}; \\[3mm] a_{21} = \dfrac{K_1 V_i - \frac{V_i^2 R_L}{V_o} - K_2 V_o}{K_1 V_o C - C V_i R_L}; & a_{22} = \dfrac{V_i - K_2 K \beta V_o}{K_1 V_o C - C V_i R_L}; \\[3mm] a_{23} = \dfrac{K_3 V_o}{K_1 V_o C - C V_i R_L}; & a_{24} = \dfrac{K_3 K \beta V_o}{K_1 V_o C - C V_i R_L}; \\[3mm] a_{31} = 1; & a_{32} = 0; \\[1mm] a_{33} = 0; & a_{34} = 0; \\[1mm] a_{41} = 0; & a_{42} = 1; \\[1mm] a_{43} = 0; & a_{34} = 0. \end{cases} \tag{11.29}$$

The above derivation has been performed with $v_i = V_i$, $r_L = R_L$, $V_{ref} - \beta V_o = 0$, and $I_{ref} = I_L = K(V_{ref} - \beta V_o)$, and the assumptions $K \gg 1$, $I_L \gg \tilde{i}_L$ and $V_o \gg \tilde{v}_o$. Therefore, the characteristic equation of the linearized system is given by

$$\begin{vmatrix} s - a_{11} & -a_{12} & -a_{13} & -a_{14} \\ -a_{21} & s - a_{22} & -a_{23} & -a_{24} \\ -1 & 0 & s & 0 \\ 0 & -1 & 0 & s \end{vmatrix} = s^4 + p_1 s^3 + p_2 s^2 + p_3 s + p_4 = 0 \tag{11.30}$$

where

$$\begin{cases} p_1 = -a_{11} - a_{22} \\ p_2 = a_{11}a_{22} - a_{12}a_{21} - a_{13} - a_{24} \\ p_3 = a_{11}a_{24} + a_{13}a_{22} - a_{12}a_{23} - a_{14}a_{21}a_{14} \\ p_4 = a_{13}a_{24} - a_{14}a_{23} \end{cases}. \tag{11.31}$$

TABLE 11.1

Specifications of the buck converter.

Description	Parameter	Nominal Value
Input voltage	v_i	24 V
Capacitance	C	150 μF
Capacitor ESR	r_C	21 mΩ
Inductance	L	100 μH
Inductor resistance	r_L	0.12 Ω
Switching frequency	f_S	20 kHz
Minimum load resistance	$r_{L(min)}$	0.75 Ω
Maximum load resistance	$r_{L(max)}$	3 Ω
Desired output voltage	V_{od}	12 V

Application of the Routh-Hurwitz criterion to the characteristic equation (11.30) shows that the system is stable if the following conditions are satisfied:

$$\begin{cases} p_1 > 0 \\ p_2 > \dfrac{p_3}{p_1} \\ p_3 > \dfrac{p_1 p_4}{p_2 - \frac{p_3}{p_1}} \\ p_4 > 0 \end{cases} \qquad (11.32)$$

Hence, by numerically solving equation (11.29) and substituting its solutions into (11.32), the stability of the system can be determined.

11.5 Results and Discussions

The application of the DISM controller on power converters has been evaluated for various topologies. In this section, computer simulation results of the PWM-based DISM buck converter and experimental results of the PWM-based DISM boost converter are provided for validation.

11.5.1 Simulation Result of PWM-Based DISM Buck Converter

The performance of the PWM-based DISM voltage controlled buck converter shown in Fig. 11.2 is compared to the PWM-based ISM voltage controlled buck

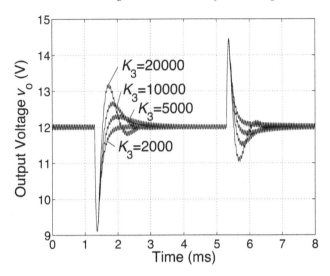

FIGURE 11.6
Output voltage waveforms of the PWM-based DISM voltage controlled buck converter operating at step load changes alternating between 3 Ω and 0.75 Ω for various values of K_3.

converter (given in Chapter 8) through computer simulations.[4] The specifications of the converter are given in Table 11.1.

Here, the PWM-based ISM controller is designed to give a critically-damped response with a bandwidth of $f_{BW} = 2.5$ kHz. Specifically, the sliding coefficients are $\frac{\alpha_1}{\alpha_2} = 4\pi f_{BW} = 31415.93$ s^{-1} and $\frac{\alpha_3}{\alpha_2} = 4\pi^2 f_{BW}^2 = 246740110$ s^{-2} [90]. The reference voltage is set as $V_{ref} = 2.5$ V, which gives $\beta = \frac{V_{ref}}{V_{od}} = 0.208$ V/V. Designing for the maximum load resistance, the control parameters are determined as $K_1 = \beta L \left(\frac{\alpha_1}{\alpha_2} - \frac{1}{r_L C} \right) = 0.608$ Hs^{-1} (or V/A) and $K_2 = \frac{\alpha_3}{\alpha_2} LC = 3.701$ V/V. Therefore, the control signal equation is

$$v_c = 0.208\, v_o - 0.608\, i_C + 3.701\,(V_{ref} - \beta v_o). \tag{11.33}$$

To compare the effectiveness of the controller, the same parameters of K_1 and K_2 used in the ISM controller are employed in the case of the DISM controller. Hence, the equation of the control signal for the PWM-based DISM controller is

$$v_c = 0.208\, v_o - 0.608\, i_C + 3.701\,(V_{ref} - \beta v_o) + K_3 \int (V_{ref} - \beta v_o)\, dt. \tag{11.34}$$

Figure 11.6 shows the output voltage waveforms of the PWM-based DISM

---

[4]The simulations are performed using Matlab/Simulink. The step size taken for all simulations is 10 ns.

voltage controlled buck converter for various values of K_3. It is worth mentioning that the steady-state voltage error is near zero for all cases of K_3. The choice of K_3 is therefore based on the required response.

Hereon, the simulations of the DISM controller are performed with the parameter $K_3 = 2000$ V/V, which is chosen for its critically-damped response. Figures 11.7(a) and 11.7(b) show the plots of the steady-state output voltage against the switching frequency of the buck converter under the PWM-based ISM and DISM controllers for respectively the maximum and minimum load resistances. As expected, the result shows that the addition of the double-integral term of the state variables does nullify the steady-state error of the converter for all switching frequencies.

Figures 11.8(a)–11.8(c) show the output voltage waveforms of the buck converter operating at 20 kHz and with the load stepping between 3 Ω and 0.75 Ω, under the PWM-based ISM controller at standard V_{ref} level, the PWM-based ISM controller with uplifted V_{ref} level, and the PWM-based DISM controller at standard V_{ref} level, respectively. Figure 11.8(a) shows the typical output voltage waveform expected in a low-switching-frequency ISM voltage controlled buck converter. The output voltage is poorly regulated with a high steady-state error, i.e., $v_o = 10.4$ V at minimum load resistance and $v_o = 10.7$ V at maximum load resistance. As mentioned before, in practice, it is possible to raise the reference setpoint to obtain the desired output voltage. This is demonstrated in Fig. 11.8(b), where V_{ref} is raised from the original value of 2.5 V to 2.78 V. Here, it can be seen that at maximum load resistance, the output voltage is regulated at 12 V. However, at minimum load resistance, the output voltage is only 11.7 V. This verifies the inadequacy of the method of reference-voltage-shifting correction in eliminating the steady-state error of the converter. Finally, it is illustrated in Fig. 11.8(c) that by converting the ISM controller into a DISM controller and with a proper choice of parameter K_3, it is possible to derive a PWM-based (indirect) type of SM controller that achieves near perfect regulation even at a low switching frequency, and that has similarly good dynamical behavior as the ISM controller.

11.5.2 Experimental Result of PWM-Based DISM Boost Converter

The performance of the PWM-based DISM current controlled boost converter shown in Fig. 11.3 is experimentally compared to the PWM-based ISM current controlled boost converter [96]. The specifications of the converter are given in Table 11.2.

Here, the PWM-based ISM controller is optimally tuned using the approach described in [96] to give the fastest critically-damped response. The reference voltage is fixed at $V_{\text{ref}} = 8$ V, which makes $\beta = G_s = 0.167$ V/V. The implemented control signal equation is

$$v_c = 4.167\left(V_{\text{ref}} - 0.167v_o\right) - i_C - 0.77i_L + 0.167\left(v_o - v_i\right). \qquad (11.35)$$

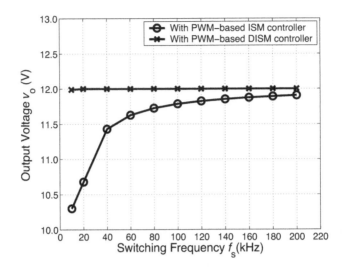

(a) At maximum load resistance $r_L = 3\ \Omega$

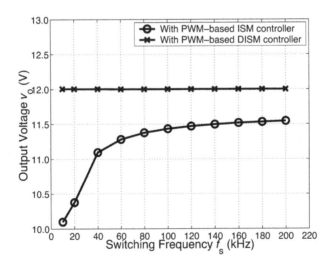

(b) At minimum load resistance $r_L = 0.75\ \Omega$

FIGURE 11.7
Plots of steady-state output voltage v_o against switching frequency f_s of the buck converter operating under the PWM-based ISM and DISM controllers at (a) maximum and (b) minimum load resistances.

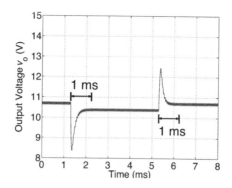

(a) ISM controller with $V_{\mathrm{ref}} = 2.5$ V

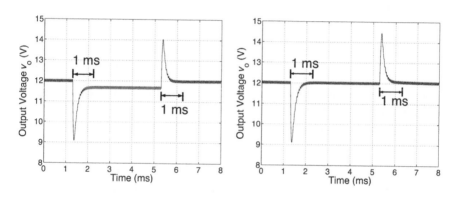

(b) ISM controller with $V_{\mathrm{ref}} = 2.78$ V   (c) DISM controller with $V_{\mathrm{ref}} = 2.5$ V

FIGURE 11.8
Output voltage waveforms of a 24 V input, 12 V output buck converter operating at step load changes alternating between 3 Ω and 0.75 Ω under (a) a PWM-based ISM controller at standard V_{ref} level; (b) a PWM-based ISM controller with uplifted V_{ref} level; and (c) a PWM-based DISM controller at standard V_{ref} level.

Next, the PWM-based DISM controller is also tuned to give the fastest critically-damped response. Here, the integral term $K_2 \int [K\,(V_{\mathrm{ref}} - \beta v_{\mathrm{o}}) - i_L]\, dt$ of the original equation (11.18) is ignored to simplify circuit implementation. This is possible since the inductor current error is negligible due to the high current-error gain value of K.[5] The final form of the control signal equation

---

[5] In the case where the gain value is small or that a large inductor current error is present, the integral term is required.

TABLE 11.2
Specifications of the boost converter.

Description	Parameter	Nominal Value
Input voltage	v_i	24 V
Capacitance	C	220 μF
Capacitor ESR	r_C	25 mΩ
Inductance	L	300 μH
Inductor resistance	r_L	0.14 Ω
Switching frequency	f_S	200 kHz
Minimum output current	$i_{r(min)}$	0.5 A
Maximum output current	$i_{r(max)}$	2 A
Desired output voltage	V_{od}	48 V

adopted in the experiment is

$$v_c = 1.58\left(V_{ref} - 0.167v_o\right) - i_C + K_2 \int \left(V_{ref} - \beta v_o\right) dt$$
$$-0.77i_L + 0.167\left(v_o - v_i\right). \qquad (11.36)$$

A comparison between (11.35) and (11.36) shows that with the additional integral term of $K_2 \int \left(V_{ref} - \beta v_o\right) dt$, the gain value of $V_{ref} - 0.167v_o$ has been lowered to 1.58 V/V in the DISM controller. The reduction in the proportional gain is to tackle the instability due to the introduction of the integral pole term.

Figure 11.9 shows the output voltage waveforms of the PWM-based DISM voltage controlled boost converter for various gains of the integral term, i.e., $K_2 = 720$ V/V, 1220 V/V, and 1930 V/V. Notably, the steady-state DC voltage error is eliminated for all three cases. Also, the variations in the dynamic behavior due to the respective K_2 value can be observed from the figure. Here, the experiments of the DISM controller are performed with the parameter $K_2 = 1220$ V/V, which is chosen for its critically damped response.

Figures 11.10(a) and 11.10(b) show the plots of the steady-state output voltage against the switching frequency of the boost converter under the PWM-based ISM and DISM controllers for respectively the minimum and maximum output currents. According to the measured results, the addition of the double-integral term in the PWM ISM controller reduces the steady-state regulation error (between minimum and maximum output currents) from $< 1\%$ to $< 0.05\%$ for all values of switching frequency. Hence, the strength of the solution in eliminating the steady-state error of the PWM ISM controller is illustrated.

Figures 11.11(a) and 11.11(b) show the output voltage waveforms of the boost converter operating at 200 kHz and step load changes alternating between 0.5 A and 2 A, under the PWM-based ISM controller and the PWM-based DISM controller, respectively. As seen from the figures, both controllers displayed excellent large-signal property (a major feature of the SM control)

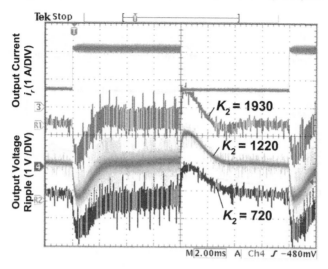

FIGURE 11.9
Output voltage waveforms of the PWM-based DISM voltage controlled boost converter operating at step load changes alternating between 0.5 A and 2 A for various values of integral capacitors.

of having a consistent response for both step up and down load changes. However, with the PWM-based ISM controller, the converter contains a significant level of steady-state error of around 400 mV (see Fig. 11.11(a)). Such error is not present with the PWM-based DISM controller (see Fig. 11.11(b)). Yet, it can be seen that with the PWM-based ISM controller, a faster dynamical response can be achieved. This is due to the higher value of the voltage-error control gain in the PWM-based ISM controller (gain value of 4.167) than the PWM-based DISM controller (gain value of 1.58) (refer to (11.35) and (11.36)). Hence, it is important to note that even though the steady-state error of the PWM-based ISM controller can be eliminated to achieve near perfect regulation by converting it into the DISM controller, it *may* come at the expense of a lower allowable proportional gain value, which gives a slower dynamical response. Otherwise, in systems in which the same gain value can be employed for both the PWM-based ISM and DISM controllers, similar dynamical behavior can be obtained with both controllers, as illustrated in the example with the buck converter.

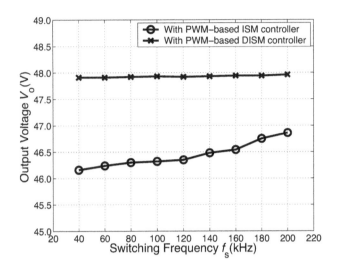

(a) At minimum output current $i_r = 0.5$ A

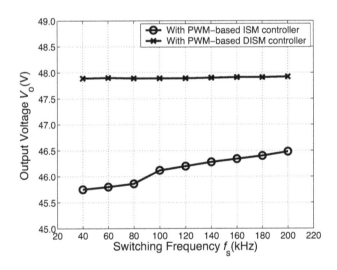

(b) At maximum output current $i_r = 2.0$ A

FIGURE 11.10
Plots of steady-state output voltage v_o against switching frequency f_s of the boost converter operating under the PWM-based ISM and DISM controllers at (a) minimum and (b) maximum output currents.

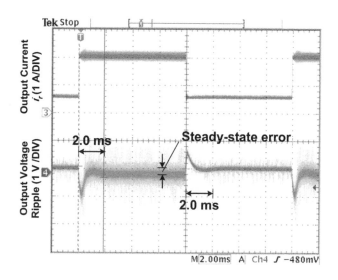

(a) ISM controller

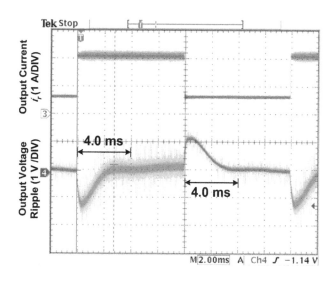

(b) DISM controller

FIGURE 11.11
Output voltage waveforms of a 24 V input, 48 V output boost converter
operating at step load changes alternating between 0.5 A and 2 A under (a)
the PWM-based ISM controller; and (b) the PWM-based DISM controller.

Bibliography

[1] J. Ackermann and V. Utkin, Sliding mode control design based on Ackermann's formula, *IEEE Transactions on Automatic Control*, vol. 43 no. 2, pp. 234–237, 1998.

[2] M. Ahmed, M. Kuisma, K. Tolsa, and P. Silventoinen, Implementing sliding mode control for buck converter, *IEEE Power Electronics Specialists Conference Record*, vol. 2, pp. 634–637, 2003.

[3] M. Ahmed, M. Kuisma, P. Silventoinen, and O. Pyrhonen, Effect of implementing sliding mode control on the dynamic behavior and robustness of switch mode power supply (buck converter), *Proceedings of Fifth International Conference on Power Electronics and Drive Systems*, vol. 2, pp. 1364–1368, 2003.

[4] M. Ahmed, M. Kuisma, O. Pyrhonen, and P. Silventoinen, Sliding mode control for buck-boost converter using control desk dSPACE, *Proceedings of Fifth International Conference on Power Electronics and Drive Systems*, vol. 2, pp. 1491–1494, 2003.

[5] E. Alarcon, A. Romero , A. Poveda, S. Porta, and L. Martinez-Salamero, Sliding-mode control analog integrated circuit for switching DC–DC power converters, *Proceedings of IEEE International Symposium on Circuits and Systems*, pp. 500–503, 2001.

[6] F. Bilalović, O. Mušić, and A. Šabanović, Buck converter regulator operating in the sliding mode, *Proceedings of Seventh International Conference on Power Conversion*, pp. 331–340, 1983.

[7] S. A. Bock, J. R. Pinheiro, H. Grundling, H. L. Hey, and H. Pinheiro, Existence and stability of sliding modes in bi-directional DC–DC converters, *IEEE Power Electronics Specialists Conference Record*, vol. 3, pp. 1277–1282, 2001.

[8] J. Calvente, L. Martinez-Salamero, and R. Giral, Design of locally stable sliding modes in bidirectional switching converters, *Proceedings of 40th Midwest Symposium on Circuits and Systems*, vol. 1, pp. 615–618, 1997.

[9] B. J. Cardoso, A. F. Moreira, B. R. Menezes, and P. C. Cortizo, Analysis of switching frequency reduction methods applied to sliding mode

controlled DC–DC converters, *Proceedings of IEEE Applied Power Electronics Conference and Exposition*, pp. 403–410, 1992.

[10] J. M. Carrasco, J. M. Quero, F. P. Ridao, M. A. Perales, and L. G. Franquelo, Sliding mode control of a DC/DC PWM converter with PFC implemented by neural networks, *IEEE Transactions on Circuits and Systems Part I: Fundamental Theory and Applications*, vol. 44 no. 8, pp. 743–749, Aug. 1997.

[11] M. Castilla, L. C. de Vicuna, M. Lopez, O. Lopez, and J. Matas, On the design of sliding mode control schemes for quantum resonant converters, *IEEE Transactions on Power Electronics*, vol. 15 no. 15, pp. 960–973, 2000.

[12] M. Castilla, L. G. de Vicuna, J. M. Guerrero, J. Matas, and J. Miret, Sliding-mode control of quantum series-parallel resonant converters via input-output linearization, *IEEE Transactions on Industrial Electronics*, vol. 52, no. 2, pp. 566–575, Apr. 2005.

[13] C. Y. Chan, A nonlinear control for DC–DC power converters, *IEEE Transactions on Power Electronics*, vol. 22, no. 1, pp. 216–222, Jan. 2007.

[14] H. C. Chan, K. T. Chau, and C. C. Chan, A neural network controller for switching power converters, *IEEE Power Electronics Specialists Conference Record*, pp. 887–892, 1993.

[15] K. H Cheng, C. F. Hsu, C. M. Lin, T. T. Lee, and C. Li, Fuzzy-neural sliding-mode control for DC–DC converters using asymmetric gaussian membership functions, *IEEE Transactions on Industrial Electronics*, vol. 54, no. 3, pp. 1528–1536, Jun. 2007.

[16] H. Chiacchiarini, P. Mandolesi, and A. Oliva, Nonlinear analog controller for a buck converter: Theory and experimental results, *Proceedings of IEEE International Symposium on Industrial Electronics*, pp. 601–606, 1999.

[17] D. Cortes and J. Alvarez, Robust sliding mode control for the boost converter, *Proceedings of IEEE International CIEP Power Electronics Congress*, pp. 208–212, 2002.

[18] S. Cuk and R. D. Middlebrook, A new optimum topology switching DC-to-DC converter, *IEEE Power Electronics Specialists Conference Record (PESC)*, pp. 160–179, 1977.

[19] F. Dominguez, E. Fossas, and L. Martinez-Salamero, Stability analysis of a buck converter with input filter via sliding-mode approach, *Proceedings of IEEE Conference on Industrial Electronics, Control and Instrumentations*, pp. 1438–1442, 1994.

[20] P. F. Donoso-Garcia, P. C. Cortizo, B. R. de Menezes, and M. A. Severo Mendes, Sliding mode control for current distribution in DC-to-DC converters connected in parallel, *IEEE Power Electronics Specialists Conference Record*, pp. 1513–1518, 1996.

[21] C. Edwards and S. K. Spurgeron, *Sliding Mode Control: Theory and Applications*. London, U.K.: Taylor and Francis, 1998.

[22] R. W. Erickson, S. Ćuk, and R. D. Middlebrook, Large-signals modelling and analysis of switching regulators, *IEEE Power Electronics Specialists Conference Record*, pp. 240–250, 1982.

[23] G. Escobar, R. Ortega, H. Sira-Ramirez, J. P. Vilain, and I. Zein, An experimental comparison of several nonlinear controllers for power converters, *IEEE Control Systems Magazine*, vol. 19, no. 1, pp. 66–82, 1999.

[24] E. Figueres, G. Garcera, J. M. Benavent, M. Pascual, and J. A. Martinez, Adaptive two-loop voltage-mode control of DC–DC switching converters, *IEEE Transactions on Industrial Electronics*, vol. 53, no. 1., pp. 239–253, Feb. 2006.

[25] A. J. Forsyth and S. V. Mollow, Modelling and control of DC–DC converters, *IEE Power Engineering Journal*, vol. 12, no. 5, pp. 229–236, 1998.

[26] E. Fossas, L. Martínez and J. Ordinas, Sliding mode control reduces audiosusceptibility and load perturbation in the Ćuk converter, *IEEE Transactions on Circuits and Systems Part I*, vol. 39, no. 10, pp. 847–849, 1992.

[27] E. Fossas and D. Biel, A sliding mode approach to robust generation on DC-to-DC nonlinear converters, *Proceedings of IEEE International Workshop on Variable Structure Systems*, pp. 67–71, 1996.

[28] E. Fossas and A. Pas, Second order sliding mode control of a buck converter, Proceedings of 41st IEEE Conference on Decision and Control, vol. 1, pp. 346–347, 2002.

[29] R. Giral, L. Martinez-Salamero, R. Leyva, and J. Maixe, Sliding-mode control of interleaved boost converters, *IEEE Transactions on Circuits and Systems Part I*, vol. 47, no. 9, pp. 1330–1339, 2000.

[30] P. Gupta, and A. Patra, Hybrid sliding mode control of DC–DC power converter circuits, *Proceedings of IEEE Region Ten Conference on Convergent Technologies for Asia-Pacific Region*, vol. 1, pp. 259–263, 2003.

[31] P. Gupta and A. Patra, Hybrid mode switched control of DC–DC boost converter circuits, *IEEE Transactions on Circuits and Systems II*, vol. 52, no. 11, pp. 734–738, Nov. 2005.

[32] D. C. Hamill, J. H. B. Deane, and D. J. Jefferies, Modelling of chaotic DC/DC converters by iterative nonlinear mappings, *IEEE Transactions on Power Electronics*, vol. 7, pp. 25–36, 1992.

[33] F. A. Himmelstoss, J. W. Kolar and F. C. Zach, Analysis of a Smith-predictor-based-control concept eliminating the right-half plane zero of continuous mode boost and buck-boost DC/DC converters, *Proceedings, International Conference on Industrial Electronics, Control and Instrumentation IECON*, pp. 423–428, Nov. 1991.

[34] S. P. Huang, H. Q. Xu, and Y. F. Liu, Sliding-mode controlled Ćuk switching regulator with fast response and first-order dynamic characteristic, *IEEE Power Electronics Specialists Conference Record (PESC)*, pp. 124–129, 1989.

[35] L. Iannelli and F. Vasca, Dithering for sliding mode control of DC/DC converters, *IEEE Power Electronics Specialists Conference Record*, vol. 2, pp. 1616–1620, 2004.

[36] J. G. Kassakian, M. F. Schlecht, and G. C. Verghese, *Principles of Power Electronics*. Reading, MA: Addison-Wesley, 1992.

[37] F. C. Lee, R. P. Iwens, Y. Yu, and J. E. Triner, Generalized computer aided discrete time domain modelling and analysis of DC–DC converters, *IEEE Transactions on Industrial Electronics and Control Instrumentation*, vol. 26, pp. 58–69, 1979.

[38] R. Leyva, L. Martinez-Salamero, B. Jammes, J. C. Marpinard, and F. Guinjoan, Identification and control of power converters by means of neural networks, *IEEE Transactions on Circuits and Systems Part I*, vol. 44, no. 8, pp. 735–742, 1997.

[39] K. H. Liu and F. C. Lee, Topological constraints on basic PWM converters, *IEEE Power Electronics Specialists Conference Record*, pp. 160–179, 1988.

[40] Y. F. Liu and P. C. Sen, A general unified large signal model for current programmed DC-to-DC converters, *IEEE Transactions on Power Electronics*, vol. 9, pp. 414–424, 1994.

[41] M. López, L. G. De-Vicuña, M. Castilla, and J. Majo, Interleaving of parallel DC-DC converters using sliding mode control, *Proceedings of IEEE Conference on Industrial Electronics, Control and Instrumentations*, pp. 1055–1059, 1998.

[42] M. López, L. G. De-Vicuña, M. Castilla, P. Gaya, and O. López, Current distribution control design for paralleled DC/DC converters using sliding-mode control, *IEEE Transactions on Industrial Electronics*, vol. 45, no. 10, pp. 1091–1100, 2004.

[43] J. Mahdavi and A. Emadi, Sliding-mode control of PWM Ćuk converter, *Proceedings of Sixth International Conference on Power Electronics and Variable Speed Drives*, vol. 2, pp. 372–377, 1996.

[44] J. Mahdavi, A. Emadi, and H. A. Toliyat, Application of state space averaging method to sliding mode control of PWM DC/DC converters, *Proceedings of IEEE Conference on Industry Applications*, vol. 2, pp. 820–827, 1997.

[45] J. Mahdavi, M.R. Nasiri, A. Agah, and A. Emadi, Application of neural networks and state-space averaging to DC/DC PWM converters in sliding-mode operation, *IEEE/ASME Transactions on Mechatronics*, vol. 10, no. 1, pp. 60–67, Feb. 2005.

[46] L. Malesani, L. Rossetto, G. Spiazzi, and P. Tenti, Performance optimization of Ćuk converters by sliding-mode control, *IEEE Transactions on Power Electronics*, vol. 10 no. 3, pp. 302–309, 1995.

[47] R. Mammano, Switching power supply topology: voltage mode vs. current mode, *Unitrode Design Note*, Jun. 1994.

[48] L. Martinez-Salamero, A. Poveda, J. Majo, L. Garcia-de-Vicuna, F. Guinjoan, J. C. Marpinard, and M. Valentin, Lie algebras modelling of bidirectional switching converters, *Proceedings of European Conference on Circuit Theory and Design*, vol. 2, pp. 1425–1429, 1993.

[49] L. Martinez-Salamero, J. Calvente, R. Giral, A. Poveda, and E. Fossas, Analysis of a bidirectional coupled-inductor Ćuk converter operating in sliding mode, *IEEE Transactions on Circuits and Systems Part I*, vol. 45, no. 4, pp. 355–363, 1998.

[50] L. Martinez-Salamero, H. Valderrama-Blavi, and R. Giral, Self-oscillating DC-to-DC switching converters with transformers characteristics, *IEEE Transactions on Aerospace and Electronic Systems*, vol. 41, no. 2, pp. 710–716, Apr. 2005.

[51] J. Matas, L. G. De-Vicuña, O. López, and M. López, Discrete sliding mode control of a boost converter for output voltage tracking, *Proceedings of Eighth International Conference on Power Electronics and Variable Speed Drives*, pp. 351–354, 2000.

[52] P. Mattavelli, L. Rossetto, G. Spiazzi, and P. Tenti, General-purpose sliding-mode controller for DC/DC converter applications, *IEEE Power Electronics Specialists Conference Record*, pp. 609–615, 1993.

[53] P. Mattavelli, L. Rossetto, and G. Spiazzi, Small-signal analysis of DC–DC converters with sliding mode control, *IEEE Transactions on Power Electronics*, vol. 12 no. 1, pp. 96–102, 1997.

[54] P. Mattavelli, L. Rossetto, G. Spiazzi, and P. Tenti, General purpose fuzzy controller for DC–DC converters, *IEEE Transactions on Power Electronics*, vol. 12, pp. 79–86, 1997.

[55] S. K. Mazumder, A. H. Nayfeh, A. Borojevic, Robust control of parallel DC–DC buck converters by combining integral-variable-structure and multiple-sliding-surface control schemes, *IEEE Transactions on Power Electronics*, vol. 17, no. 3, pp. 428–437, 2002.

[56] S. K. Mazumder and S. L. Kamisetty, Experimental validation of a novel multiphase nonlinear VRM controller, *IEEE Power Electronics Specialists Conference Record (PESC)*, pp. 2114–2120, June 2004.

[57] R. D. Middlebrook and S. Ćuk, A general unified approach to modeling switching power converter stages, *IEEE Power Electronics Specialists Conference Record*, pp. 18–34, 1976.

[58] D. M. Mitchell, *DC–DC Switching Regulator Analysis*. New York: McGraw Hill, 1998.

[59] C. Morel, J. C. Guignard, and M. Guillet, Sliding mode control of DC-to-DC power converters, *Proceedings of 9th International Conference on Electronics, Circuits and Systems*, vol. 3, pp. 971–974, 2002.

[60] C. Morel, Slide mode control via current mode control in DC–DC converters, *Proceedings of IEEE International Conference on Systems, Man and Cybernetics*, vol. 5, pp. 6–11, 2002.

[61] C. Morel, Application of slide mode control to a current-mode-controlled boost converter, *Proceedings, IEEE Conference on Industrial Electronics, Control and Instrumentations*, vol. 3, pp. 1824–1829, 2002.

[62] V. M. Nguyen and C. Q. Lee, Tracking control of buck converter using sliding-mode with adaptive hysteresis, *IEEE Power Electronics Specialists Conference Record*, vol. 2, pp. 1086–1093, 1995.

[63] V. M. Nguyen and C. Q. Lee, Indirect implementations of sliding-mode control law in buck-type converters, *Proceedings of IEEE Applied Power Electronics Conference and Exposition*, vol. 1, pp. 111–115, 1996.

[64] A. Ofoli and A. Rubaai, Real-time implementation of a fuzzy logic controller for switch-mode power-stage DC–DC converters, *IEEE Transactions on Industry Applications*, vol. 42, no. 6, pp. 1367–1374, Nov./Dec. 2006.

[65] K. Ogata, *Modern Control Engineering*. Upper Saddle River, NJ: Prentice-Hall, Inc., 1997.

[66] M. Oppenheimer, M. Husain, M. Elbuluk, and J. A. Garcia, Sliding mode control of the Ćuk converter, *IEEE Power Electronics Specialists Conference Record*, vol. 2, pp. 1519–1526, 1996.

[67] R. Orosco and N. Vazquez, Discrete sliding mode control for DC/DC converters, *Proceedings of IEEE International CIEP Power Electronics Congress*, pp. 231–236, 2000.

[68] W. Perruquetti and J. P. Barbot, *Sliding Mode Control in Engineering.* New York: Marcel Dekker, 2002.

[69] A. G. Perry, F. Guang, Y. F Liu, and P. C. Sen, A new sliding mode like control method for buck converter, *IEEE Power Electronics Specialists Conference Record*, vol. 5, pp. 3688–3693, 2004.

[70] V. S. C. Raviraj and P. C. Sen, Comparative study of proportional-integral, sliding mode, and fuzzy logic controllers for power converters, *IEEE Transactions on Industry Applications*, vol. 33, no. 2, pp. 518–524, 1997.

[71] R. Redl and N. O. Sokal, Current-mode control, five different types, used with the three basic classes of power converter: Small-signal AC and large-signal DC characterization, stability requirements, and implementation of practical circuits, *IEEE Power Electronics Specialists Conference Record*, pp. 771–785, 1985.

[72] R. Ridley, Current mode or voltage mode? *Switching Power Magazine*, pp. 4–9, Oct. 2000.

[73] D. J. Shortt and F. C. Lee, Extensions of the discrete-average models for converter power stages, *IEEE Power Electronics Specialists Conference Record*, pp. 23–37, 1983.

[74] Y. B. Shtessel, A. S. I. Zinober, and I. A. Shkolnikov, Boost and buck-boost power converters control via sliding modes using method of stable system centre, *Proceedings of 41st IEEE Conference on Decision and Control*, vol. 1, pp. 340–345, 2002.

[75] Y. B. Shtessel, A. S. I. Zinober, and I. A. Shkolnikov, Boost and buck-boost power converters control via sliding modes using dynamic sliding manifold, *Proceedings of 41st IEEE Conference on Decision and Control*, vol. 3, pp. 2456–2461, 2002.

[76] Y. B. Shtessel, O. A. Raznopolov, and L. A. Ozerov, Sliding mode control of multiple modular DC-to-DC power converters, in *Proceedings of IEEE International Conference on Control Applications*, pp. 685–690, 1996.

[77] Y. B. Shtessel, O. A. Raznopolov, and L. A. Ozerov, Control of multiple modular DC-to-DC power converters in conventional and dynamic sliding surfaces, *IEEE Transactions on Circuits and Systems Part I*, vol. 45, no. 10, pp. 1091–1100, 1998.

[78] H. Sira-Ramirez and M. Ilic, A geometric approach to the feedback control of switch mode DC-to-DC power supplies, *IEEE Transactions on Circuits and Systems*, vol. 35 no. 10, pp. 1291–1298, 1988.

[79] H. Sira-Ramirez, A geometric approach to pulse-width modulated control in nolinear dynamical systems, *IEEE Transactions on Automatic Control*, vol. 34 no. 3, pp. 184–187, 1989.

[80] H. Sira-Ramirez and M. Rios-Bolivar, Sliding mode control of DC-to-DC power converters via extended linearization, *IEEE Transactions on Circuits and Systems Part I*, vol. 41 no. 10, pp. 652–661, 1994.

[81] H. Sira-Ramirez, R. Ortega, R. Perez-Moreno, and M. Garcia-Esteban, A sliding mode controller-observer for DC-to-DC power converters: A passivity approach, *Proceedings of 34th IEEE Conference on Decision and Control*, pp. 3379–3384, 1995.

[82] H. Sira-Ramirez, G. Escobar, and R. Ortega, On passivity-based sliding mode control of switched DC-to-DC power converters, *Proceedings of 35th IEEE Conference on Decision and Control*, pp. 2525–2526, 1996.

[83] H. Sira-Ramirez, Sliding mode-Δ modulation control of a buck converter, *Proceedings of 42nd IEEE Conference on Decision and Control*, vol. 3, pp. 2999–3004, 2003.

[84] H. Sira-Ramirez, On the generalized PI sliding mode control of DC-to-DC power converters: A tutorial, *International Journal of Control*, vol. 76 no. 9/10, pp. 1018–1033, 2003.

[85] J. J. E. Slotine and W. Li, Chapter 7: Sliding control, in *Applied Nonlinear Control*. Englewood Cliffs, NJ: Prentice-Hall, Inc., 1991.

[86] K. M. Smedley and S. Ćuk, One-cycle control of switching converters, *IEEE Transactions on Power Electronics*, vol. 10, pp. 625–633, 1995.

[87] W. C. So, C. K. Tse and Y. S. Lee, Development of a fuzzy logic controller for DC/DC converters: Design, computer simulation and experimental evaluation, *IEEE Transactions on Power Electronics*, vol. 11, pp. 24–32, 1996.

[88] G. Spiazzi and P. Mattavelli, Chapter 8: Sliding-mode control of switched-mode power supplies, in *The Power Electronics Handbook*, T. L. Skvarenina (Editor), Boca Raton, FL: CRC Press LLC, 2002.

[89] S. C. Tan, Y. M. Lai, M. K. H. Cheung, and C. K. Tse, An adaptive sliding mode controller for buck converter in continuous conduction mode, *Proceedings, IEEE Applied Power Electronics Conference and Exposition*, pp. 1395–1400, 2004.

[90] S. C. Tan, Y. M. Lai, M. K. H. Cheung, and C. K. Tse, On the practical design of a sliding mode voltage controlled buck converter, *IEEE Transactions on Power Electronics*, vol. 20, no. 2, pp. 425–437, 2005.

[91] S. C. Tan, Y. M. Lai, C. K. Tse, and M. K. H. Cheung, A fixed-frequency pulse-width-modulation based quasi-sliding mode controller for buck converters, *IEEE Transactions on Power Electronics*, vol. 20, no. 6, pp. 1379–1392, 2005.

[92] S. C. Tan, Y. M. Lai, and C. K. Tse, Implementation of pulse-width-modulation based sliding mode controller for boost converters, *IEEE Power Electronics Letters*, vol. 3, no. 4, pp. 130–135, 2005.

[93] S. C. Tan, Y. M. Lai, C. K. Tse, and M. K. H. Cheung, Adaptive feed-forward and feedback control schemes for sliding mode controlled power converters, *IEEE Transactions on Power Electronics*, vol. 21, no. 1, pp. 182–192, 2006.

[94] S. C. Tan, Y. M. Lai, and C. K. Tse, A unified approach to the design of PWM based sliding mode voltage controller for DC–DC converters in continuous conduction mode, *IEEE Transactions on Circuits and Systems I*, vol. 53, no. 8, pp. 1816–1827, 2006.

[95] S. C. Tan, Y. M. Lai, C. K. Tse, and L. Martinez-Salamero, Special family of PWM-based sliding mode voltage controllers for basic DC-DC converters in discontinuous conduction mode, *IET Electric Power Applications*, vol. 1, no. 1, pp. 64–74, Jan. 2007.

[96] S.C. Tan, Y.M. Lai, C.K. Tse, and C.K. Wu, Indirect sliding mode control of power converters via double integral sliding surface, *IEEE Transactions on Power Electronics*, vol. 23, no. 2, pp. 600–611, Mar. 2008.

[97] C. K. Tse and K. M. Adams, Qualitative analysis and control of a DC-DC boost converter operating in discontinuous mode, *IEEE Transactions on Power Electronics*, vol. 5, pp. 323–330, 1990.

[98] C. K. Tse and K. M. Adams, An adaptive control for DC-DC converters, *IEEE Power Electronics Specialists Conference Record (PESC)*, pp. 213–218, 1990.

[99] C. K. Tse and K. M. Adams, A nonlinear large-signal feedforward-feedback control for two-state DC-DC Converters, *IEEE Power Electronics Specialists Conference Record*, pp. 722–729, 1991.

[100] C. K. Tse and K. M. Adams, Quasi-linear analysis and control of DC-DC Converters, *IEEE Transactions on Power Electronics*, vol. 7, pp. 315–323, 1992.

[101] V. Utkin, *Sliding Modes in Control Optimization*. Berlin: Springer-Verlag, 1992.

[102] V. Utkin, J. Guldner, and J. X. Shi , *Sliding Mode Control in Electromechanical Systems*. London, U.K.: Taylor and Francis, 1999.

[103] N. Vazquez, C. Hernandez, J. Alvarez, and J. Arau, Sliding mode control for DC/DC converters: A new sliding surface, *Proceedings of IEEE International Symposium on Industrial Electronics*, vol. 1, pp. 422–426, 2003.

[104] R. Venkataramanan, A. Šabanović, and S. Ćuk, Sliding mode control of DC-to-DC converters, *Proceedings of IEEE Conference on Industrial Electronics, Control and Instrumentations*, pp. 251–258, 1985.

[105] R. Venkataramanan, *Sliding Mode Control of Power Converters*, PhD thesis, California Institute of Technology, California, 1986.

[106] G. C. Verghese, M. E. Elbuluk, and J. G. Kassakian, A general approach to sample-data modelling for power electronics circuits, *IEEE Transactions on Power Electronics*, vol. 1, pp. 74–89, 1986.

[107] G. C. Verghese, C. A. Bruzos, and K. N. Mahabir, Averaged and sampled-data models for current mode control: A re-examination, *IEEE Power Electronics Specialists Conference Record*, pp. 484–491, 1989.

[108] E. Vidal-Idiarte, L. Martinez-Salamero, F. Guinjoan, J. Calvente, and S. Gomariz, Sliding and fuzzy control of a boost converter using an 8-bit microcontroller, *IEE Proceedings—Electric Power Applications*, vol. 151, no. 1, pp. 5–11, Jan. 2004.

[109] E. Vidal-Idiarte, L. Martinez-Salamero, J. Calvente, and A. Romero, An H_∞ control strategy for switching converters in sliding-mode current control, *IEEE Transactions on Power Electronics*, vol. 21, pp. 553–556, Mar. 2006.

[110] K. Viswanathan, R. Oruganti, and D. Srinivasan, Nonlinear function controller: a simple alternative to fuzzy logic controller for a power electronic converter, *IEEE Transactions on Industrial Electronics*, vol. 52, no. 5, pp. 1439–1448, Oct. 2005.

[111] H. W. Whittington, B. W. Flynn, and D. E. Macpherson, *Switched Mode Power Supplies: Design and Construction*. New York: Wiley, 2nd ed., 1997.

[112] C. C. Wu and C. M. Young, A new PWM control strategy for the buck converter, *Proceedings, IEEE Conference on Industrial Electronics Society (IECON)*, pp. 157–162, 1999.

Index

282

Printed and bound by CPI Group (UK) Ltd, Croydon, CR0 4YY

18/10/2024

01776264-0009